AF576005

Polymers and Ionic Liquids

Polymers and Ionic Liquids: Shaping up a New Generation of High Performances Nanomaterials

Editors

Jannick Duchet-Rumeau
Jean-François Gérard
Sébastien Livi

MDPI • Basel • Beijing • Wuhan • Barcelona • Belgrade • Manchester • Tokyo • Cluj • Tianjin

Editors
Jannick Duchet-Rumeau
Laboratoire Ingénierie des Matériaux Polymères
France

Jean-François Gérard
Laboratoire Ingénierie des Matériaux Polymères
France

Sébastien Livi
Laboratoire Ingénierie des Matériaux Polymères
France

Editorial Office
MDPI
St. Alban-Anlage 66
4052 Basel, Switzerland

This is a reprint of articles from the Special Issue published online in the open access journal *Nanomaterials* (ISSN 2079-4991) (available at: https://www.mdpi.com/journal/nanomaterials/special_issues/ionic_liquid_nano).

For citation purposes, cite each article independently as indicated on the article page online and as indicated below:

LastName, A.A.; LastName, B.B.; LastName, C.C. Article Title. *Journal Name* **Year**, *Article Number*, Page Range.

ISBN 978-3-03943-308-7 (Hbk)
ISBN 978-3-03943-309-4 (PDF)

Contents

About the Editors

Jannick Duchet-Rumeau completed her Ph.D. at the University of Lyon in 1996, where she focused on the modelling of the interface in a polyethylene/glass system, tuned by the grafting connecting chains to improve the adhesion properties. She held a post-Ph.D. position at Louvain La Neuve (Belgium), where she worked on polymer nanotubes, in 1998. She started working as a full professor in 2010 at the laboratory 'Ingénierie des Matériaux Polymères' UMR CNRS #5223. Her research topics are related to nanocomposites materials, mesoporous materials, nanomaterials from ionic liquids (Head of GDR LIPS http://www.gdr-lips.fr) and the tailoring of interfaces/interphases in the heterogeneous materials.

Jean-François Gérard, after leading the research laboratory 'Ingénierie des Matériaux Polymères', being Vice-President for Research of INSA Lyon from 2011 to 2016, and after having been Deputy Managing Director of the transfer company SATT PULSALYS, Prof. Gerard is currently Deputy Scientific Director of the CNRS Institute of Chemistry in Paris. He obtained his Ph.D. in Polymer Science on the synthesis of zwitterionic polyurethanes based on sulfobetainic diols for self-emulsifying glass fiber sizing. He joined CNRS as a permanent scientist on the designing of interfaces and interphases in multicomponent polymer-based materials. His expertise deals with: hybrid organic/inorganic nanomaterials, nanocomposites based on thermoset and thermoplastics with a special attention to the interfacial phenomena and nanostructuration/self-assembling processes. He also acts as Vice-President of the European Center for Nanostructured Polymers.

Sébastien Livi obtained his Ph.D. in Macromolecular Materials and Composites from INSA Lyon in 2010 on the use of ionic liquids as multifunctional agents of polymer matrices. After a postdoctoral research position at Cornell University with Prof. Emmanuel P. Giannelis on the surface treatment of nanoparticles under supercritical CO2, he obtained, in 2011, the position of assistant professor at the laboratory 'Ingénierie des Matériaux Polymères' of INSA Lyon UMR CNRS #5223. His current research interests include the synthesis of ionic liquids, the structure/properties relationships of polymer nanocomposites, and polymer-IL nanomaterials based on thermoplastics and thermosets, as well as the processing of polymer foams.

Preface to "Polymers and Ionic Liquids: Shaping up a New Generation of High Performances Nanomaterials"

In the past few years, the scientific community, including in academic and industrial research, has made unprecedented efforts to develop new functional polymeric nanomaterials, in terms of water or gas barriers, electrical, mechanical, fire retardancy, or self-healing properties. In order to achieve this, the introduction of organic–inorganic hybrids, based on silica, carbon nanotubes and layered silicates, or ionomers and block copolymers, have been widely reported in the literature. Very recently, ionic liquids that are organic salts have demonstrated their great potential as new components of advanced polymer materials. In fact, they can be used as interfacial agents of nanoparticles to develop high performance nanocomposites, as compatibilizing agents of polymer blends, as electrolytes in batteries, as flame retardant of polymer materials, as structuration and self-healing agents of thermoplastic and thermosets, and as processing aids of renewable resources. Thus, the main objective of this Special Issue is to highlight a glimpse of the real potential of ionic liquids in polymer nanoscience.

Jannick Duchet-Rumeau, Jean-François Gérard, Sébastien Livi
Editors

Article

Ionic Liquid-Nanostructured Poly(Methyl Methacrylate)

Clarice Fedosse Zornio, Sébastien Livi *, Jannick Duchet-Rumeau and Jean-François Gerard *

Ingénierie des Matériaux Polymères, Université de Lyon, CNRS, UMR 5223, INSA Lyon, F-69621 Villeurbanne, France

* Correspondence: sebastien.livi@insa-lyon.fr (S.L.); jean-francois.gerard@insa-lyon.fr (J.-F.G.); Tel.: +33-472438291 (S.L.)

Received: 4 September 2019; Accepted: 21 September 2019; Published: 26 September 2019

Abstract: Here, ionic liquids (ILs) based on imidazolium and ammonium cations were used as modifying agents for poly(methyl methacrylate) (PMMA) by extrusion. The effects of the chemical nature of the cation and/or counter anion on the resulting properties of IL-modified PMMA blends were analyzed. It was found that the use of low amounts of ILs (2 wt.%) improved the thermal stability. A plasticizing effect of ILs is evidenced by a decrease in glass transition temperature Tg of the modified PMMA, allowing to get large strains at break (i.e., up to 280% or 400%) compared to neat PMMA. The deformation and fracture mechanisms of PMMA under uniaxial tensile stress (i.e., crazing) reveal that the presence of IL delayed the strain during the initiation step of crazing.

Keywords: ionic liquids; PMMA; plasticizer; mechanical behavior; crazing; thermal behavior

---

1. Introduction

In the field of engineering polymer materials, poly(methyl methacrylate) (PMMA) is a polymer material that is considered in many applications because of its low density, high optical clarity, high rigidity, easiness for processing, as well as its excellent chemical and ultraviolet resistance [1,2]. Thus, PMMA has been integrated in numerous applications such as medical devices, glass replacement, automotive parts, and so on [3–9]. However, PMMA is a brittle polymer material that has a low fracture toughness, which limits its uses [10]. In order to enhance its properties, different routes have been investigated: (i) the incorporation of inorganic particles such as silica, alumina, layered silicates, calcium carbonate, or carbon nanotubes [11–14]; (ii) the use of a dispersed rubber phase [15–17] or the blending with other thermoplastic polymers such as polyethylene (PE) and polycarbonate (PC) [18]; and (iii) the introduction of low molar mass components that act as plasticizers [19,20]. The addition of silica nanoparticles is a very promising route to enhance the thermal and mechanical properties of PMMA [21,22]. Different authors highlighted that the incorporation of silica nanoparticles could lead to a significant improvement of fracture toughness of the PMMA matrix [23,24].

Recently, ionic liquids (ILs) have appeared as new additives in polymers, allowing the design of advanced polymer-based materials [25]. ILs are organic salts that exhibit a low melting temperature, good ionic conductivity, good thermal stability, and negligible vapor pressure [26–28]. In fact, ILs were commonly used as suitable compounds for compatible agents of polymer blends [29–32], such as (nano)structuration agents of fluorinated polymers and copolymers [33,34], non-conventional initiators of epoxy reactive systems [35,36], as well as surfactants or plasticizers of different polymers [37–40]. Recent works reported the advantages ILs used as plasticizers in synthetic and biosourced polymers. The ability to act as plasticizers for imidazolium-based ILs was also reported for starch. In fact, 1-butyl-3-methylimidazolium chloride ((C_4mIm)(Cl)) IL contributes in screening the numerous hydrogen bonds leading to a significant decrease of the glass transition temperature, Tg, and acts as a processing aid in the molten state of thermoplastic starch [38]. Others authors have investigated the influence of trihexyl(tetradecyl)phosphonium bis(trifluoromethylsulfonylimide) (($P_{6,6,6,14}$)(TFSI)) and 1-pentyl-3-methylimidazolium hexafluorophosphate ((C_5mIm)(PF_6)) ILs in poly(vinyl chloride) (PVC) [41] and poly(lactic acid) (PLA) [42], respectively. Rahman et al. also demonstrated that ILs based on ammonium, imidazolium, and phosphonium cations could be an alternative to traditional plasticizers of PVC, as such compounds have a diffusion rate (i.e., better leaching and migration resistance) [37]. Concerning amorphous polymers being glassy at room temperature, Scott et al. demonstrated the ability of imidazolium-based ILs to induce plastic deformation in PMMA. In fact, one of the conventional plasticizer agents used in PMMA (i.e., dioctyl phthalate, DOP) has a molecular structure close to the imidazolium cation, (i.e., an aromatic ring with saturated alkyl chains) [43,44].

As a consequence, the main goal of this work is to study the effect of a small quantity (2 wt.%) of ILs on the physical properties of IL-modified PMMA. Moreover, the influence of the chemical nature of the cation and/or anion on the morphology, thermal stability, and surface properties as well as the mechanical properties of PMMA have been studied. Deeper attention was brought for the first time, to our knowledge, to the crazing mechanism in PMMA under uniaxial tensile stress in order to identify the basic mechanisms involved in the large strain mechanical properties of IL-modified PMMA.

2. Materials and Methods

2.1. Materials and Characterization Methods

PMMA was provided by Oroglas (Arkema Group). The molar masses were determined by Size-exclusion chromatography (SEC) with values of 30,300 and 36,000 g·mol^{-1}, respectively. Imidazolium and ammonium-based ILs were supplied by Solvionic Co. and were denoted as 1-butyl-3-methylimidazolium hexafluorophosphate ((C_4mIm)(PF_6)), N-trimethyl-N-butylammonium hexafluorophosphate (($N_{1,1,1,4}$)(PF_6)), N-trimethyl-N-hexylammonium bromide (($N_{1,1,1,6}$)(Br)), and N-trimethyl-N-hexylammonium bis(trifluoromethanesulfonyl)imide (($N_{1,1,1,6}$)(TFSI)). All the structures of ILs and PMMA are given in Figure 1. In addition, the designation of ILs as well as their melting temperatures, T_m, and molar masses, M_m, are reported in Table 1.

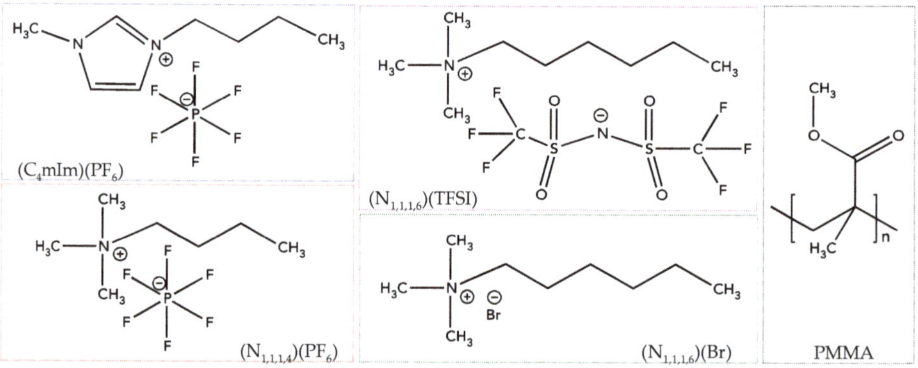

(C$_4$mIm)(PF$_6$) (N$_{1,1,1,6}$)(TFSI) (N$_{1,1,1,4}$)(PF$_6$) (N$_{1,1,1,6}$)(Br) PMMA

Figure 1. Chemical structures of ionic liquids (ILs) considered in this study. PMMA, poly(methyl methacrylate); TFSI, bis(trifluoromethanesulfonyl)imide; PF_6, hexafluorophosphate; Br, Bromide.

Table 1. Designation of imidazolium and ammonium ILs used in this work.

Designation	Cation	Anion	T_m (K)	M_m ($g{\cdot}mol^{-1}$)
$(C_4mIm)(PF_6)$	1-butyl-3-methylimidazolium	Hexafluorophosphate	283	284
$(N_{1,1,1,4})(PF_6)$	N-trimethyl-N-butylammoniun	Hexafluorophosphate	393	289
$(N_{1,1,1,6})(Br)$	N-trimethyl-N-hexylammonium	Bromide	268	224
$(N_{1,1,1,6})(TFSI)$	N-trimethyl-N-hexylammonium	Bis(trifluoromethanesulfonyl)imide	308	424

Fourier transform infrared spectroscopy (FTIR) spectra were recorded at 25 °C using a Nicolet iSO10 ThermoScientific spectrometer equipped with an Attenuated Total Reflectance (ATR) mode setup from 4000 to 600 cm^{-1}.

Differential scanning calorimetry (DSC) analyses were performed using a Q20 TA instrument (TA Co. Ltd., New Castle, DE, USA). The samples were investigated from 193 to 453 K at a heating rate of 10 $K{\cdot}min^{-1}$ under N_2 flow (50 $mL{\cdot}min^{-1}$). Before analysis, the samples were heated to 393 K to erase the thermal history. From DSC measurements, the glass transition temperature, Tg, was determined.

Thermogravimetric analysis (TGA) was performed using a Q500 thermogravimetric analyzer from TA Instruments. All samples were heated from 298 to 1173 K at different heating rates under inert atmosphere. T_{onset}, T_{max}, and T_{final} temperatures were determined. The plots ln(β/T2) versus 1/T obtained from TGA traces for each mass degradation and heating rate provided the activation energy of the degradation mechanism [45]. The degradation kinetics were analyzed according to the Kissinger–Akahira–Sunose method (KAS) equation [46].

$$\ln\left(\frac{\beta}{T^2}\right) = \text{Constant} - \frac{E_a}{RT},$$

where β is the heating rate (in $K{\cdot}min^{-1}$), T is the temperature (in K) recorded at each degree of degradation (defined as the fraction of the total mass loss in the process, ranging from 10 to 90%), R is the universal gas constant, and E_a is the activation energy (in $kJ{\cdot}mol^{-1}$).

Transmission electron microscopy (TEM) was performed at the Technical Center of Microstructures, at University of Lyon, using a Phillips CM 120 field emission scanning electron microscope (Philips, Amsterdam, The Netherlands) with an accelerating voltage of 80 kV. First, the samples were sliced using an ultramicrotome (Leica, Weitzlar, Germany) equipped with a diamond knife to obtain ultrathin sections 60 nm thick. Then, they were set on cooper grids for observation.

The surface energy of PMMA/IL materials was determined from the sessile drop method using a DataPhysics Instruments (GmbH) OCA 20 (Filderstadt, Germany). Water and diiodomethane were used as probe liquids for contact angle measurements. The nondispersive and dispersive components of surface energy were determined according to the Owens–Wendt model [47].

Uniaxial tensile tests were performed on IL-modified PMMA samples in order to determine the Young's modulus and the strain at break. The dumbbell-shaped specimens were tested using an MTS 2/M eletromechanical testing system (MTS, Eden Prairie, MN, USA) at 295 K under 50% relative humidity (RH) with a crosshead speed of 2 $mm{\cdot}min^{-1}$. In addition, in order to analyze more deeply the crazing mechanism, the samples were strained under a uniaxial tensile stress at different levels of strain (i.e., ranging from 20% to 80%) to be observed by optical microscopy (OM).

2.2. *Processing of IL-Modified PMMA*

PMMA was combined with 2 wt.% of ionic liquid ($(C_4mIm)(PF_6)$, $(N_{1,1,1,4})(PF_6)$, (N1,1,1,6)(Br), or (N1,1,1,6)(TFSI)) in molten state using a 15 g capacity microextruder (DSM 15) with co-rotating twin screws. PMMA/IL mixtures were processed with the following parameters: 5 min at 100 rpm at a temperature of 180 °C and injected at 90 °C in order to generate samples for mechanical tests.

3. Results & Discussion

3.1. PMMA/IL Interactions

The miscibility of ILs in the PMMA matrix after melt mixing was examined by TEM. The TEM micrographs of neat PMMA and IL-modified PMMA materials are shown in Figure 2. The first conclusion is that the evidenced morphologies were in the nanometer range, which can explain why the IL-modified PMMA blends kept the initial PMMA optical properties.

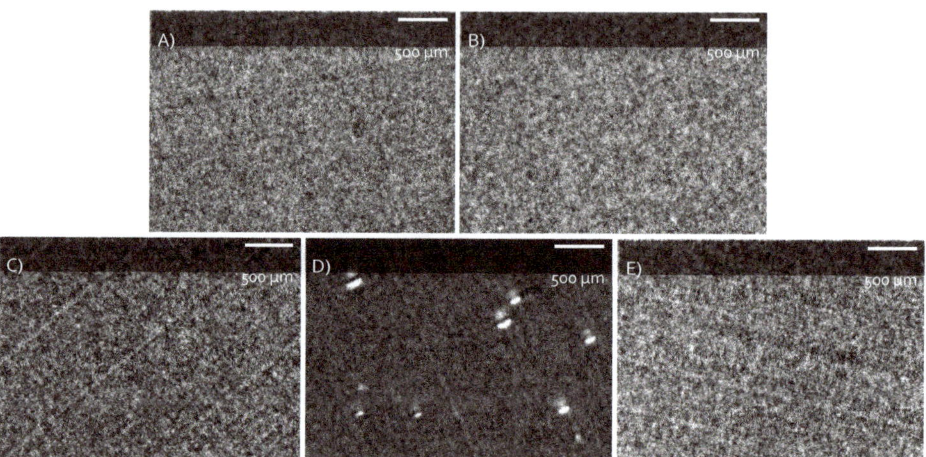

Figure 2. TEM micrographs of neat PMMA and IL-modified PMMA materials: (**A**) neat PMMA; (**B**) PMMA/(C_4mIm)(PF_6); (**C**) PMMA/($N_{1,1,1,4}$)(PF_6); (**D**) PMMA/($N_{1,1,1,6}$)(Br); and (**E**) PMMA/($N_{1,1,1,6}$)(TFSI) (2 wt.% IL).

TEM micrographs did not show large differences between PMMA, PMMA/(C_4mIm)PF_6), PMMA/($N_{1,1,1,4}$)PF_6), and PMMA/($N_{1,1,1,6}$)(TFSI) morphologies. Such a phenomenon can be explained by the excellent miscibility of such imidazolium and ammonium ILs into the PMMA matrix [48]. On the other hand, one can notice on TEM micrographs of PMMA/($N_{1,1,1,6}$)(Br) the presence of tiny voids (with about 0.5 µm in diameter) scarcely dispersed in the PMMA matrix. This suggested a poor miscibility of the ($N_{1,1,1,6}$)(Br) ionic liquid in PMMA.

To provide a better understanding of the type of intermolecular interactions between PMMA and imidazolium or ammonium-based ILs, FTIR spectroscopy was used. FTIR spectra of ILs, PMMA, and the different IL-modified PMMA materials are given in Figure 3.

The ammonium and imidazolium ionic liquids were well characterized by the absorption bands at 2939 and 2878 cm^{-1}, attributed to the C–H stretching vibration, and by the bands at 1467 and 751 cm^{-1}, corresponding to the C=N and C–N stretching, respectively (Figure 3A). The symmetric as well as asymmetric stretching of the $(PF_6)^-$ anion appeared at 819 and 740 cm^{-1}, respectively [41]. For ($N_{1,1,1,4}$)(PF6), the absorption bands associated to the anion shifted to 823 and 738 cm^{-1}. Despite the different lengths of the alkyl chains of ammonium cations (i.e., $(N_{1,1,1,4})^+$ and $(N_{1,1,1,6})^+$), the characteristic absorption bands were similar (Figure 3): C–H stretching was evidenced in the 3050–2860 cm^{-1} range; C–H bending at 1490–1480 cm^{-1}; and at the lowest wavelengths, 970 and 900 cm^{-1}, the characteristic vibrations of the C–N–C torsion bonds occurred. Concerning ($N_{1,1,1,6}$)(Br) IL, the broad absorption band from 3600 to 3000 cm^{-1} was related to the water uptake, due the more pronounced hydrophilic character of this IL, which is associated to the $(Br)^-$ anion. On the other hand, for the ($N_{1,1,1,6}$)(TFSI) IL, the strong hydrophobicity of the $(TFSI)^-$ anion prevents water uptake, and there was no evidence of residual water. The same conclusion could be made for $(PF_6)^-$ -based

ILs [49]. The $(TFSI)^-$ anion was characterized by strong absorption bands between 740 and 1350 cm^{-1}: at 1347, 1328, and 1133 cm^{-1} for the SO_2 group; from 1226 to 1178 cm^{-1} for the $–CF_3$ group; and at 1052 and 788 cm^{-1} for S–N–S and C–S vibrations, respectively [50,51].

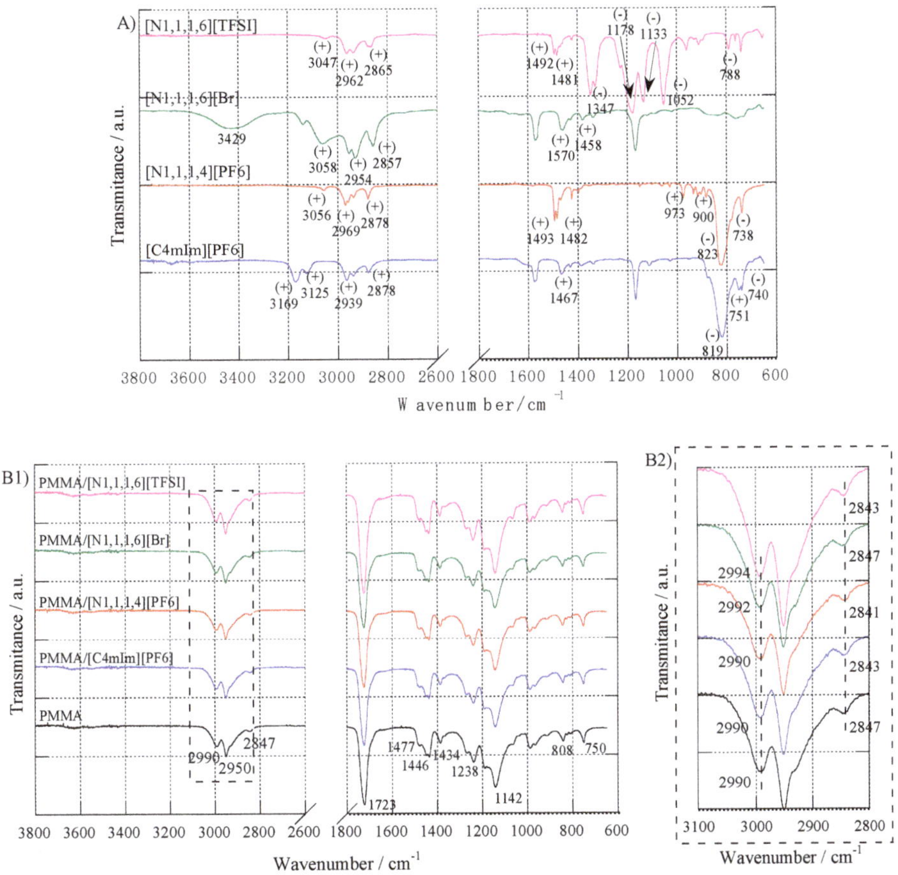

Figure 3. FTIR spectra of: (**A**) neat imidazolium and ammonium-based ILs, (+) and (−) captions highlight the main absorption bands for the cation and anion, respectively; (**B1**) neat PMMA and IL-modified PMMA materials; and (**B2**) displays the same spectra ranging only from 3100 to 2800 cm^{-1}).

Figure 3B reports the spectra of PMMA and IL-modified PMMA materials. For PMMA, absorption bands appearing at 2990 and 2847 cm^{-1} corresponded to the asymmetric and symmetric stretches of C–H in the CH_3 group, respectively. At 2950 cm^{-1}, the $–CH_2–$ groups of the polymer backbone displayed their asymmetric stretching vibration mode. The C=O vibration modes were evidenced at 1723 (stretching), 808 (in-plane banding), and 750 cm^{-1} (out of-plane bending). The $–CH_3$ deformation modes (1477, 1446, and 1434 cm^{-1}), twisting (1142 cm^{-1}), and the C–O bond stretching mode at 1238 cm^{-1} were also present.

For FTIR spectra of all IL-modified PMMA materials, one can notice that only the vibrational bands assigned to PMMA were observed according to the very small amount of IL added to PMMA. Nevertheless, it is possible for IL-modified materials to evidence shifts in the absorption bands related to methyl stretching vibrations (from 3100 to 2800 cm^{-1}—see Figure 3B2). Thus, whatever the IL used, strong interactions between PMMA and IL took place.

In fact, according to the PMMA structure, the negative charge delocalized between the two oxygens of the methacrylate group induced a positive charge on the hydrogens of the methyl group. As a consequence, the $(PF_6)^-$, $(Br)^-$, and $(TFSI)^-$ anions can interact with the $-OCH_3$ groups of PMMA (especially for $(PF_6)^-$ and $(TFSI)^-$ because of the strong electronegativity of fluorine). As a consequence, one can consider that the compatibility between PMMA and the IL is dictated by the nature of the anion rather than the cation. In fact, this conclusion was indicated by Ueno et al. who analyzed the solubility of PMMA in several ILs. The authors demonstrated that the interactions between the polymer and ILs were governed by the anion structure, whereas the cation had secondary effects [52].

As reported in Figure 3B2 for PMMA/$(N_{1,1,1,6})$(TFSI), both C–H symmetric and asymmetric stretching absorption bands shifted by 4 cm^{-1} compared to the ones of neat PMMA. On the other side, for PMMA combined with $(C_4mIm)(PF_6)$ or $(N_{1,1,1,4})(PF_6)$, only a shift of the C–H symmetric vibrational band could be evidenced (i.e., from 2847 cm^{-1} for neat PMMA to 2843 cm^{-1} for PMMA/$(C_4mIm)(PF_6)$, and 2841 cm^{-1} for PMMA/$(N_{1,1,1,4})(PF_6)$). No significant difference could be noticed between FTIR spectra of neat PMMA and PMMA/$(N_{1,1,1,6})$(Br). From these observations, one can conclude that $(TFSI)^-$ displayed higher interactions with PMMA, whereas no specific interactions were created between $(N_{1,1,1,6})$(Br) and PMMA. These conclusions are in agreement with the literature, which reports that PMMA and ILs interact via van der Waals forces, and that $(TFSI)^-$ exhibits the highest affinity with PMMA among a series of anions according to its high hydrophobicity [52–54].

Differential scanning calorimetry was also carried out to analyze the consequences of these intermolecular interactions between PMMA and ionic liquids on the Tg of PMMA (Table 2).

Table 2. Effect of imidazolium- or ammonium-based ILs on the Tg (DSC) and thermal stability—T_{onset}, T_{max}, and T_{final} corresponding to the starting, maximum, and end of degradation—considering the first derivative of the weight loss vs. temperature, respectively.

Material	T_g/K	T_{onset}/K	T_{max}/K	T_{final}/K
PMMA	367	614	656	691
PMMA/$(C_4mIm)(PF_6)$	360	627	667	708
PMMA/$(N_{1,1,1,4})(PF_6)$	364	617	662	705
PMMA/$(N_{1,1,1,6})$(Br)	363	502 [a]/614 [b]	548 [a]/659 [b]	693
PMMA/$(N_{1,1,1,6})$(TFSI)	367	619	666	699

[a,b] correspond to values obtained for the first and second step of degradation, respectively.

The Tg of PMMA remains the same after the addition of $(N_{1,1,1,6})$(TFSI) ionic liquid, whereas a decrease was noticed for the other IL-modified PMMA materials. As expected from the conclusions on the poor interactions of $(N_{1,1,1,6})$(TFSI) with PMMA, no plasticization occurred [24]. In the opposite, the incorporation of ILs such as $(N_{1,1,1,4})$(PF6) and $(N_{1,1,1,6})(Br)^-$ into PMMA led to a slight decrease of the Tg (4 °C). Comparing the cation structures, it is evident that the acid hydrogen of the imidazolium ring (N = CH − N) can interact with the carboxyl group, enhancing the compatibility between the IL and polymer matrix. On the other hand, the Tg for PMMA/$(C_4mIm)(PF_6)$ decreased about 7 °C compared to PMMA because of the specific interactions between PMMA and both $(C_4mIm)^+$ and $(PF_6)^-$, which compete with the intermolecular interactions between the polymer chains. As previously mentioned, many works described the use of ILs as plasticizing agents of polymer materials, especially because of their similar characteristics to conventional plasticizers, such as low volatility, low leachability, high temperature stability, and compatibility with polymers [43,55]. Moreover, ideal plasticizers possess a solvating group located internally rather than as terminal groups. Thereby, the charged structures of ILs are not much different [37].

Based on FTIR and DSC analyses, the respective interactions between PMMA and ILs can be summarized as reported in Figure 4. (As no interactions between PMMA and ammonium cations occur, the structures of $(N_{1,1,1,4})^+$ and $(N_{1,1,1,6})^+$ for $(N_{1,1,1,4})(PF_6)$ and $(N_{1,1,1,6})$(TFSI) ionic liquids are not represented.) On the other hand, $(N_{1,1,1,6})(Br)$ is represented as a ion pair to highlight its poor interactions with PMMA.

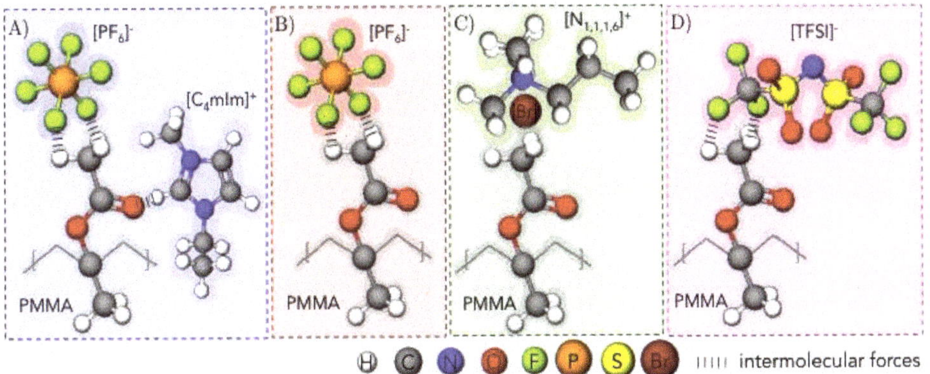

Figure 4. Schematic representations of the interactions between the methyl group of PMMA ($-OCH_3$) and (**A**) $(C_4mIm)(PF_6)$, (**B**) $(N_{1,1,1,4})(PF_6)$, (**C**) $(N_{1,1,1,6})(Br)$, and (**D**) $(N_{1,1,1,6})(TFSI)$ ILs.

3.2. *Consequences of the Presence of ILs on the Thermal Stability of PMMA*

The thermal stability, determined by TGA, of neat ILs, neat PMMA, and IL-modified PMMA materials is shown in Figure 5.

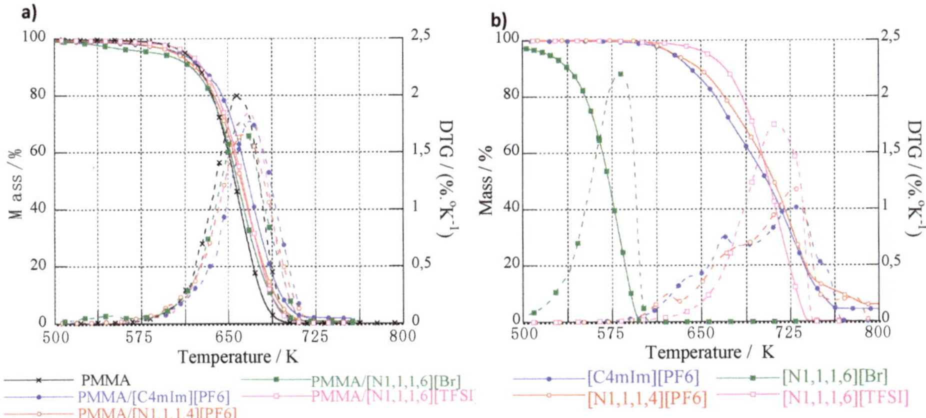

Figure 5. TGA (**a**) and DTG (first derivative) (**b**) traces of imidazolium- and ammonium-based ILs and IL-modified PMMA materials.

3.2.1. Thermal Stability of ILs

Among ILs, $(N_{1,1,1,6})(TFSI)$ displayed the better thermal stability (i.e., higher than $(C_4mIm)(PF_6)$ and $(N_{1,1,1,4})(PF_6)$). The thermal stabilities of the counter anions can be ranked in the following order: $(TFSI)^- > (PF_6)^- >> (Br)^-$, which is in agreement with the literature. In fact, it is well known that the thermal stability of ILs is primarily dependent on the hydrophobicity of the anion, which could be associated with its nucleophilicity [56,57]. Indeed, the thermal decomposition of ILs have been attributed to the decomposition of the cation induced by nucleophilic attack of the anion [58–60]. Thus, the poor nucleophilicity of $(TFSI)^-$ and $(PF_6)^-$ confers a high thermal stability to the IL [61].

In addition, other works highlighted that the thermal degradation of ILs is mainly controlled by the chemical nature of the counter anion [62,63]. Our results show a decrease of 13 K of the T_{onset} of

$(N_{1,1,1,4})(PF_6)$ compared to $(C_4mIm)(PF_6)$, which demonstrates clearly the effect of the nature of the cation. Usually, imidazolium cations are considered more thermally stable than the corresponding ammonium ones. Such an effect can be related to the fact that the thermal treatment of imidazolium cations induces rearrangements in the 1-substitute imidazole groups to prevent ring scission, whereas ammonium cations undergo Hofmann's elimination under heating [61,64,65].

It is also important to notice that several works reported, as a general feature, that the longer the alkyl chain length, the less thermally stable the IL, as a longer alkyl chain confers an increased stability to the carbocation [56]. However, our results do not show any evidence about the effect of the alkyl chain of the cation on the thermal properties of ILs. $(N_{1,1,1,6})(TFSI)$, having the longer alkyl chain, is the most stable compound. Such evidence contradicts the expectation that the imidazolium-based IL ($(C_4mIm)(PF_6)$) is more thermally stable than the ammonium ones. Overall, a clear effect of the nature of the anion on the thermal stability is observed for such ILs.

3.2.2. Thermal Stability of IL-Modified PMMA Materials

The lowest values for T_{max}, T_{onset}, and T_{final} were related to neat PMMA, and improved thermal stability was achieved for IL-modified PMMA materials (Table 3). PMMA/$(C_4mIm)(PF_6)$ displayed the highest thermal stability compared to PMMA/$(N_{1,1,1,4})(PF_6)$ and PMMA/$(N_{1,1,1,6})(TFSI)$. The main information obtained from TGA analyses for neat PMMA and IL-modified PMMA materials was the good compatibility between PMMA and $(C_4mIm)(PF_6)$, $(N_{1,1,1,4})(PF_6)$, and $(N_{1,1,1,6})(TFSI)$, as a single-step of degradation was evidenced. In the opposite, the use of $(N_{1,1,1,6})(Br)$ as a modifying agent for PMMA led to the appearance of an additional degradation peak at lower temperatures (about 553 K). In fact, this first step occurred in the same temperature range for the thermal decomposition of neat $(N_{1,1,1,6})(Br)$. Such a phenomenon evidences the nonmiscibility of this IL with PMMA and confirms the conclusions issued from TEM analyses.

3.3. *Surface Properties of IL-Modified PMMA*

The influence of the introduction of ILs on the surface properties of IL-modified PMMA materials was studied by a sessile drop method. The data values obtained for neat PMMA and IL-modified PMMA materials are summarized in Table 3.

Table 3. Determination of nondispersive and dispersive components of the surface energy of neat PMMA and IL-modified PMMA materials from contact angles with water and diiodomethane.

Sample	θ_{H2O} (°)	θ_{CH2I2} (°)	γ_t (mN·m^{-1})	γ^d (mN·m^{-1})	γ^{nd} (mN·m^{-1})
PMMA	71 ± 1	40.1 ± 1.7	40.7	30.3	10.4
PMMA/$(C_4mIm)(PF_6)$	72 ± 1	27.0 ± 1.1	45.3	40.1	5.2
PMMA/$(N_{1,1,1,4})(PF_6)$	72 ± 1	26.6 ± 1.7	45.3	40.5	4.8
PMMA/$(N_{1,1,1,6})(Br)$	66 ± 1	34.0 ± 0.9	44.6	35.1	9.5
PMMA/$(N_{1,1,1,6})(TFSI)$	73 ± 1	40 ± 0.7	43.0	37.8	5.2

Whatever the chemical nature of ILs, their incorporation into the PMMA matrix induced an increase in the total surface energy. The major contribution to this phenomenon was the significant increase in the dispersive component combined with a decrease in the nondispersive component for IL-modified PMMA materials. This increase in the total surface energy lies on the nature of interactions acting at the surface [66]. In fact, in the case of PMMA, only van der Waals forces were involved, whereas Coulomb forces were superimposed for IL-modified PMMA surfaces. The changes observed with the nature of the IL can be explained by their chemical structures. In fact, ILs are usually considered as moderate polar compounds due to their 3D structuration (i.e., a polar core surrounded by dispersive domains). Thus, an increase in the alkyl chain length of the cation and in the anion size leads to a decrease in the surface energy of the IL, especially because of the ion charge [67]. As shown by Santos et al., for various imidazolium-based ILs, the dispersive domain increases when the alkyl chain increases. Thus, the number of van der Waals interactions between ion pairs increases, while the Coulombic forces remain constant [68]. Thus, as $(N_{1,1,1,6})^+$ has a longer alkyl chain than the butyl chains present in $(C_4mIm)^+$ and $(N_{1,1,1,4})^+$ cations, a decrease in the ratio of Coulomb to van der Waals forces is expected (i.e., a slighter decrease in the surface energy of materials prepared with $(N_{1,1,1,6})^+$-based ILs is induced). In addition, we can assume that the lowest surface energy obtained for PMMA/$(N_{1,1,1,6})$(TFSI) is due to the additional effect in this balance of intermolecular forces of the large $[TFSI]^-$ anion compared to the small $(Br)^-$ anion. The nondispersive components of all IL-modified PMMA materials were lower than the one obtained for neat PMMA, as the incorporation of ILs led to more hydrophobic materials. However, it is important to notice that the nondispersive components for PMMA/$(C_4mIm)(PF_6)$, PMMA/$(N_{1,1,1,4})(PF_6)$, and PMMA/$(N_{1,1,1,6})$(TFSI) were quite similar compared to the higher value obtained for PMMA/$(N_{1,1,1,6})$(Br). These results are in agreement with the expected phenomena, as it is widely reported in the literature that $[PF_6]^-$ and $(TFSI)^-$ could be considered as hydrophobic anions, while $(Br)^-$ is considered hydrophilic [69]. In fact, the anion could be used to tailor the hydrophobicity/hydrophilic balance of the IL, but the size of the cation could also play a role [38,39].

3.4. Mechanical Properties of IL-Modified PMMA

Figure 6 reports the mechanical curves of neat PMMA and IL-modified PMMA materials (see Table 4 summarizing the Young's modulus as well as elongation at break for the various materials).

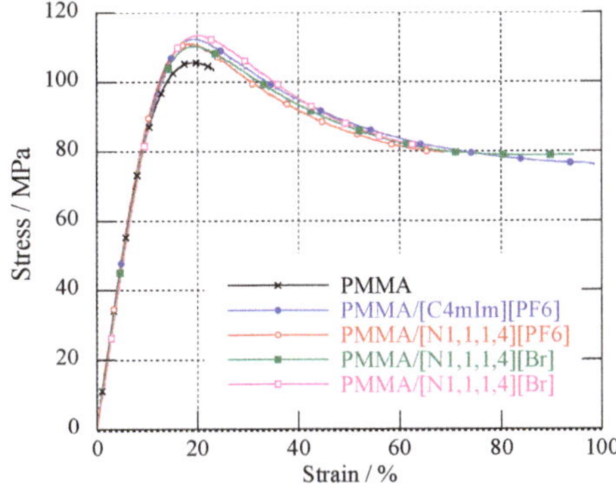

Figure 6. Strain–stress curves in uniaxial tension for neat PMMA and IL-modified PMMA materials (295 K, 50% RH, 2 mm·min^{-1}).

Table 4. Effect of ILs on the tensile properties of IL-modified PMMA at 295 K. (50% RH, 2 mm·min^{-1}).

Material	Young's modulus (MPa)	Strain at Break (%)
PMMA	953 ± 7	24 ± 1
PMMA/$(C_4mIm)(PF_6)$	939 ± 2	96 ± 3
PMMA/$(N_{1,1,1,4})(PF_6)$	943 ± 5	70 ± 4
PMMA/$(N_{1,1,1,6})$(Br)	920 ± 7	92 ± 3
PMMA/$(N_{1,1,1,6})$(TFSI)	934 ± 2	68 ± 4

PMMA is considered as a rigid and brittle polymer having a high Young's modulus and low elongation at break [1,70]. Thus, usually, plasticizers are added to the polymer in order to enhance the ability of the material to sustain large deformations and improve toughness [71]. As shown in Table 4, a slight plasticizing effect of the ILs could be observed, whereas a very large increase in the strain at break was obtained for IL-modified PMMA materials compared to neat PMMA, whatever the nature of the added ionic liquid was. In addition, one can notice that for all the IL/PMMA materials, the yield stress remained similar, whatever the ionic liquid nature was, and was slightly higher than the one of the neat PMMA.

In fact, the addition of 2 wt.% of IL did not significantly impact the Young's modulus of IL-modified PMMA. These results are similar to those reported by Scott et al., who studied PMMA plasticized with two imidazolium-based ILs at higher concentrations [43,44]. In fact, these authors reported that the Young's modulus of PMMA is the same with the addition of 10 wt.% of IL, whereas it decreases dramatically for IL contents from 20 to 50 wt.%. Nevertheless, the lowest value of Young's modulus obtained for PMMA/($N_{1,1,1,6}$)(Br) could be attributed to the poor interactions between this IL and PMMA, as reported previously. This fact leads to some exudation of the IL and voids formation.

As mentioned, the presence of ILs in PMMA strongly influences the strain at break of the material. In a similar work done for polyvinylchloride, Rahman et al. reported the ability of an ILs series to improve the flexibility of PVC [37]. In their work, the strain at break of the various IL-plasticized PVC was in the range from 2% to 94% depending on the chemical nature of IL (i.e., much higher than that of the strain at break obtained for neat PVC (1.4%)). In our work, we demonstrated very clearly that even larger improvements could be achieved with the addition of IL into PMMA without main changes to the Young's modulus. Except for PMMA/($N_{1,1,1,6}$)(Br), the extent to which the different ILs imparted plastic deformation of PMMA clearly lies with the Tg values (shown Table 2). In fact, PMMA/(C_4mIm)(PF_6) material, which has the lowest Tg, displayed the highest strain at break, whereas PMMA/($N_{1,1,1,4}$)(PF_6) and PMMA/($N_{1,1,1,6}$)(TFSI) materials, having similar Tgs, displayed a strain at break of about 280%.

According to these observations on Young's modulus and yield stress, it is obvious that the presence of IL mainly influenced the plastic deformation mechanisms (i.e., crazing- and/or shear bands-based phenomena).

3.5. *Influence of ILs on the Deformation Phenomena of IL-Modified PMMA*

In order to have a better understanding of the role of the IL on the deformation processes of IL-modified PMMA materials, the initiation and growth crazing mechanisms have been investigated The process of deformation and fracture of polymers under mechanical stress involves three main steps: the first one occurs at the molecular scale involving polymer chain re-arrangements, conformations changes, disentanglements, and scissions; the second step is governed by the appearance of crazes (crazing initiation, growth, and breakdown) at the microscale level, followed by initiation and propagation of microcracks; and the final step occurs at the macroscopic level with the material failure induced by the extension of cracks [72].

Under uniaxial tension, crazes propagate perpendicular to the direction of uniaxial stress and grow by extension and break of internal fibrils under specific conditions (see the scheme of a craze in Figure 7) [72,73]. Thus, the crazing mechanism of IL-modified PMMA materials was followed as a function of strain under uniaxial tensile stress. Figure 7 reports the optical microscopy micrographs of neat PMMA and IL-modified PMMA materials at different strain levels. In order to provide a quantitative analysis of the crazing process, the values for the average craze length and width were recorded as a function of the tensile strain (from 20% to 80% strain) (Table 5). Neat PMMA has a strain at break that is too low to sustain large tensile strains as the IL/PMMA materials. Thus, the crazes characteristics are reported only for 20% strain.

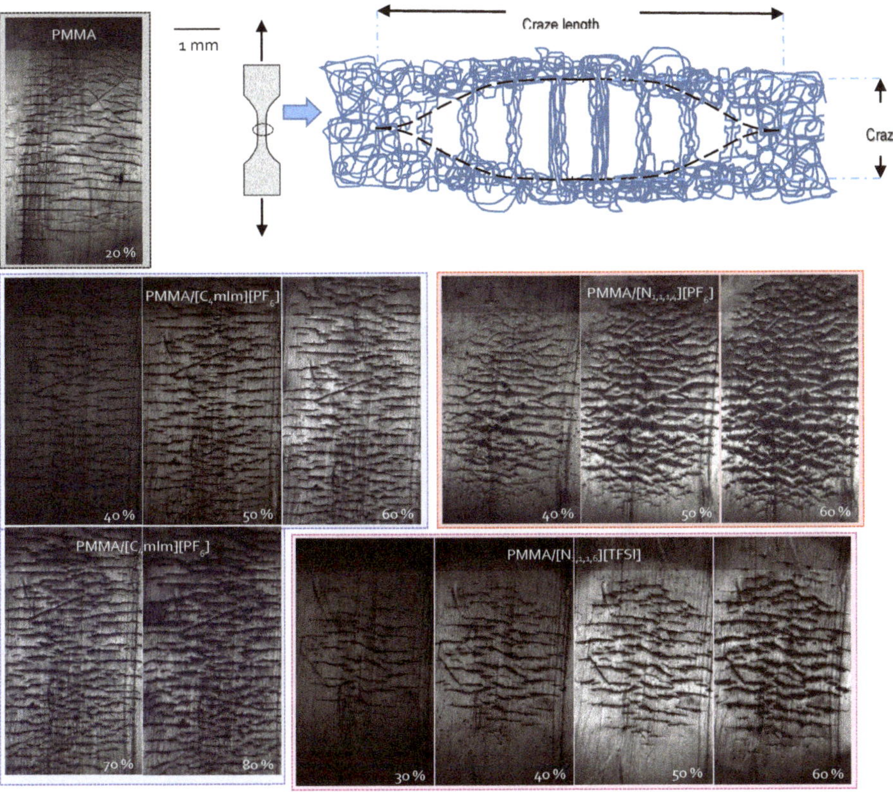

Figure 7. OM micrographs of neat PMMA and IL-modified PMMA materials at different tensile strain levels.

Table 5. Craze width, length, and density obtained for neat PMMA and IL-modified for different given strains during a uniaxial tensile test performed at 20 °C.

	Strain (%)	Craze Width (μm)	Craze Length (μm)	Craze Density (crazes·mm^{-2})
PMMA	20	256	415	770
PMMA/ (C_4mIm)(PF_6)	40	18	332	1060
	50	23	361	1030
	60	24	395	970
PMMA/ ($N_{1,1,1,4}$)(PF_6)	40	22	416	490
	50	30	499	520
	60	39	531	550
PMMA/ ($N_{1,1,1,6}$)(TFSI)	40	24	285	340
	50	31	464	340
	60	39	535	350

As observed on OM micrographs, in the case of the neat PMMA, the appearance of crazes preceded the fracture to an extremely limited extent, whereas for IL-modified PMMA materials, the crazing growth mechanism was considerably slower and depended on the chemical nature of the IL used to modify PMMA.

It is known that the appearance and growth of crazes is an efficient phenomenon for toughening thermoplastics. In fact, craze growth plays a key role by absorbing energy from surface creation and chain pull out during the deformation process [74–77]. Thus, the generation of numerous nucleation sites of crazing enhances the amount of energy absorbed and contributes efficiently to the material toughness by delaying fracture. It is also well known that by enhancing the chain mobility, usually by increasing temperature, crazing initiation and the extension of the internal fibrils from chain pull out are facilitated [74,75]. It was previously demonstrated that ILs lead to a decrease in the intermolecular interactions between polymer chains. In fact, the presence of ILs delays the strain at which the first crazes appear. Observable crazes appeared at 20% of strain for neat PMMA, whereas 30% of strain must be reached for $(N_{1,1,1,6})(Br)$- and $(N_{1,1,1,6})(TFSI)$-modified PMMA and 40% for $(C_4mIm)(PF_6)$- and $(N_{1,1,1,4})(PF_6)$-modified PMMA.

On the other hand, comparing the craze widths and lengths as a function of strain could give information on the effect of the presence of ILs on the toughening mechanisms of IL-modified PMMA. The presence of $(C_4mIm)(PF_6)$ IL clearly induced the appearance of the smallest crazes in PMMA for a given strain but also led to the slowest crazes growth mechanism. One can mention that for 60% of strain, the values for the average craze width and length of the PMMA/$(C_4mIm)(PF_6)$ blend are smaller than the those of the neat PMMA at 20% of strain. These craze parameters (craze widths and lengths) displayed intermediate values for PMMA/$(N_{1,1,1,4})(PF_6)$ and $(N_{1,1,1,6})(TFSI)$ blends but considerably large ones for the PMMA/$(N_{1,1,1,6})(Br)$ blend. In addition, the special role of the $(C_4mIm)(PF_6)$ ionic liquid could be evidenced from the craze density (i.e., crazes per mm^{-2}), compared to the neat PMMA and the other PMMA/IL blends. This peculiar behavior can be attributed not only to the more efficient plasticizing effect of $(C_4mIm)(PF_6)$ but also to its better miscibility (see Table 3). Compared to the effect of $(C_4mIm)(PF_6)$, which induces a high strain at break associated to the plasticizing effect of the imidazolium-based IL, the introduction of $(N_{1,1,1,6})(Br)$ did not induce any plasticization of PMMA/$(N_{1,1,1,6})(Br)$ according to its poorest miscibility (Table 2). Hence, the tiny voids might explain the high strain at break observed for PMMA/$(N_{1,1,1,6})(Br)$. As a consequence, the domains of $(N_{1,1,1,6})(Br)$ observed by TEM in the PMMA are supposed to act as stress concentration zones inducing a failure at lower strain. In fact, up to 70% of strain, craze lengths and widths considerably increase, whereas the craze density decreases; this phenomenon is mainly governed by the propagation and coalescence of the previously formed crazes.

As a conclusion, crazing initiation and propagation analyses clearly evidence the effect of the nature of the IL on the mechanical properties of IL-modified PMMA. For PMMA/$(C_4mIm)(PF_6)$, PMMA/$(N_{1,1,1,4})(PF_6)$, and PMMA/$(N_{1,1,1,6})(TFSI)$, a substantial increase in the strain at break was due to the plasticizing effect induced by the addition of IL, whereas for PMMA/$(N_{1,1,1,6})(Br)$, the poor miscibility leading to low cohesion $(N_{1,1,1,6})(Br)$ IL inclusions acted as stress concentration zones, which limited the strain break value of the corresponding blend.

4. Conclusions

The effect of the introduction of a low amount (2 wt.%) of imidazolium- and ammonium-based ILs in PMMA was evaluated in order to highlight the role of the cation and the anion on the final properties of the blends. Morphology analyses suggest a good miscibility between $(C_4mIm)(PF_6)$, $(N_1,1,1,4)(PF_6)$, and $(N_{1,1,1,6})(TFSI)$ ionic liquids with PMMA. On the other hand, a poor interaction between PMMA and $(N_{1,1,1},6)(Br)$ IL was observed, which can be explained by the strong cohesion between the ions of this IL in preventing interactions with the polymer. FTIR spectroscopy evidences that specific van der Waals interactions exist between the highly miscible ILs (i.e., $(C_4mIm)(PF_6)$, $(N_{1,1,1,4})(PF_6)$, and $(N_{1,1,1,6})(TFSI)$) and the PMMA chains (due to the anion of the IL and the methyl groups of PMMA). In addition, the extent of these interactions contributes to the plasticizing effect of the ILs, as demonstrated by the slight decrease of the glass transition temperature. As no change on the Tg of PMMA/$(N_{1,1,1,6})(Br)$ was noticed compared to neat PMMA, blends based on $(N_{1,1,1,4})(PF_6)$, $(N_{1,1,1,6})(TFSI)$, and$(C_4mIm)(PF_6)$ display an increase in the polymer chain mobility, as evidenced by a decrease of Tg. One can conclude that the specific interactions between PMMA and both $[C_4mIm]^+$ and $[PF_6]^-$ ions can explain the larger plasticizing effect induced by $(C_4mIm)(PF_6)$ IL on PMMA.

The stronger the intermolecular interaction between IL and PMMA, the more thermally stable the PMMA/IL blends are, as evidenced by thermal analysis. These conclusions are issued from the increase of the activation energies, E_a, of the thermal degradation, which suggests that the ILs can stabilize the radicals released during thermal degradation of PMMA. ILs positively influence the mechanical properties of the IL-modified PMMA blends. In fact, very limited decreases of the Young's modulus combined with improved strain at break were observed and could be related to the plasticizing effect imparted by imidazolium and ammonium-based ILs on PMMA, except for PMMA/($N_{1,1,1,6}$)(Br). In addition, crazing initiation and propagation investigations clarified the influence of the IL on the mechanical behavior of PMMA. For PMMA/(C_4mIm)(PF_6), PMMA/($N_{1,1,1,4}$)(PF_6), and PMMA/($N_{1,1,1,6}$)(TFSI), the large increase of the strain at break of the IL-modified PMMA could be attributed to the plasticization effect of the IL. In the opposite direction, for PMMA/($N_{1,1,1,6}$)(Br), which was evidenced by TEM to be nonmiscible in PMMA, the intermediate behavior could be attributed to the heterogeneity of the IL-modified PMMA blend, as ($N_{1,1,1,6}$)(Br)-rich domains act as stress concentration zones (i.e., defects). As a conclusion, imidazolium- and ammonium-based IL can be proposed as efficient additives of PMMA, acting as processing aids in melt extrusion, and they lead to PMMA-like materials having improved thermal and mechanical behaviors.

Author Contributions: Conceptualization, S.L., J.D.-R. and J.F.G.; Methodology, C.F.Z.; Formal Analysis, C.F.Z. and J.-F.G.; Investigation, C.F.Z.; Data Curation, all authors; Validation, all authors; Visualization, C.F.Z.; Writing—Original Draft Preparation, C.F.Z.; Writing—Review and Editing, all authors; Resources: J.-F.G.; Project Administration, S.L., J.D.-R. and J.-F.G.

Funding: The authors gratefully acknowledge the National Council for Scientific and Technological Development (CNPq)/Brazil for C.F.Z.'s PhD grant (Ciência sem Fronteiras program).

Conflicts of Interest: The authors declare no conflicts of interest.

References

1. Agrawal, S.; Patidar, D.; Dixit, M.; Sharma, K.; Saxena, N.S.; Pratap, A.; Saxena, N.S. Investigation of Thermo-Mechanical Properties of PMMA. *AIP Conf. Proc.* **2010**, *79*, 79–82.
2. Ali, U.; Karim, K.J.B.A.; Buang, N.A. A Review of the Properties and Applications of Poly (Methyl Methacrylate) (PMMA). *Polym. Rev.* **2015**, *55*, 678–705. [CrossRef]
3. Carraher, C.E., Jr. Free Radical Chain Polymerization: Addition Polymerization. In *Introduction to Polymer Chemistry*, 3rd ed.; CRC Press: Boca Raton, FL, USA, 2012.
4. DiMaio, F.R. The science of bone cement: A historical review. *Orthopedics* **2002**, *25*, 1399–1407. [PubMed]
5. Pawar, E. A Review Article on Acrylic PMMA. *IOSR J. Mech. Civ. Eng.* **2016**, *13*, 1–4.
6. Higgs, W.A.; Lucksanasombool, P.; Higgs, R.J.E.D.; Swain, M.V. Comparison of the material properties of PMMA and glass ionomer based cements for use in orthopedic surgery. *J. Mater. Sci. Mater. Med.* **2001**, *12*, 453–460. [CrossRef] [PubMed]
7. Frazer, R.Q.; Byron, R.T.; Osborne, P.B.; West, K.P. PMMA: An essential material in medicine and dentistry. *J. Long. Term. Eff. Med. Implant.* **2005**, *15*, 629–639. [CrossRef]
8. Ramakrishna, S.; Mayer, J.; Wintermantel, E.; Leong, K.W. Biomedical applications of polymer-composite materials: A review, Compos. *Sci. Technol.* **2001**, *61*, 1189. [CrossRef]
9. Ree, S.H.; Choi, J.Y. Synthesis of a Bioactive Poly (Methyl Methacrylate)/Silica Hybrid. *Key Eng. Mater.* **2002**, *218*, 433–436.
10. Tjong, S.E. Structural and mechanical properties of polymer nanocomposites. *Mater. Sci. Eng. R Rep.* **2006**, *53*, 73–197. [CrossRef]
11. Larson, W.R.; Dixon, D.L.; Aquilino, S.A.; Clancy, J.M. The effect of carbon graphite fiber reinforcentent on the strength of provisional crown and fixed partial denture resins. *J. Prosthet. Dent.* **1991**, *66*, 816–820. [CrossRef]
12. Kim, K.S.; Byun, J.H.; Lee, G.H.; Park, S.J. Influence of GMA grafted MWNTs on physical and rheological properties of PMMA-based nanocomposites by in situ polymerization. *Macromol. Res.* **2011**, *19*, 14–20. [CrossRef]

13. Wu, W.; He, T.; Chen, J.; Zhang, X.; Chen, Y. Study on in situ preparation of nano calcium carbonate/PMMA composite particles. *Mater. Lett.* **2006**, *60*, 2410–2415. [CrossRef]
14. Nikolaidis, A.K.; Achilias, D.S.; Karayannidis, G.P. Synthesis and characterization of PMMA/organomodified montmorillonite nanocomposites prepared by in situ bulk polymerization. *Ind. Eng. Chem. Res.* **2011**, *50*, 571–579. [CrossRef]
15. Saladino, M.L.; Motaung, T.E.; Luyt, A.S.; Spinella, A.; Nasillo, G.; Caponetti, E. The effect of silica nanoparticles on the morphology, mechanical properties and thermal degradation kinetics of PMMA. *Polym. Degrad. Stab.* **2012**, *97*, 452–459. [CrossRef]
16. Etienne, S.; Becker, C.; Ruch, D.; Grignard, B.; Cartigny, G.; Detrembleur, C.; Calberg, C.; Jerome, R. Effects of incorporation of modified silica nanoparticles on the mechanical and thermal properties of PMMA. *J. Therm. Anal. Calorim.* **2007**, *87*, 101–104. [CrossRef]
17. Jayasuriya, M.M.; Hourston, D.J. The Effect of Composition and the Level of Crosslinking of the Poly(methylmethacrylate) Phase on the Properties of Natural Rubber-Poly(methylmethacrylate) Semi-2 Interpenetrating Polymer Networks. *J. Appl. Polym. Sci.* **2012**, *124*, 3558–3564. [CrossRef]
18. Gutteridge, D.L. Reinforcement of poly (methyl methacrylate) with ultra-high-modulus polyethylene fibre. *J. Dent.* **1992**, *20*, 50–54. [CrossRef]
19. Duan, H.; Zhang, L.; Chen, G. Plasticizer-assisted bonding of Poly (Methyl methacrylate) microfluidic chips at low temperature. *J. Chromatogr. A* **2010**, *1217*, 160–166. [CrossRef]
20. Flora, X.H. Role of Different Plasticizers in Li-Ion Conducting Poly (Acrylonitrile)-Poly (Methyl Methacrylate) Hybrid Polymer Electrolyte. *Int. J. Polym. Mater. Polym. Biomater.* **2013**, *62*, 37–41. [CrossRef]
21. Garcia, N.; Corrales, T.; Guzman, J.; Tiemblo, P. Understanding the role of nanosilica particle surfaces in the thermal degradation of nanosilica–poly (methyl methacrylate) solution-blended nanocomposites: From low to high silica concentration. *Polym. Degrad. Stab.* **2007**, *92*, 635–643. [CrossRef]
22. Chau, J.L.H.; Hsieh, C.C.; Lin, Y.M.; Li, A.K. Preparation of transparent silica–PMMA nanocomposite hard coatings. *Prog. Org. Coat.* **2008**, *62*, 436. [CrossRef]
23. Lach, R.; Kim, G.M.; Michler, G.H.; Grellmann, W.; Albrecht, K. Indentation Fracture Mechanics for Toughness Assessment of PMMA/SiO2 Nanocomposites. *Mater. Mater. Eng.* **2006**, *291*, 263–271.
24. Chau, J.L.H.; Hsieh, C.C.; Lin, Y.M.; Li, A.K. Effect of nanoparticle size and size-distribution on mechanical behavior of filled amorphous thermoplastic polymers. *J. Appl. Polym. Sci.* **2007**, *105*, 2577–2587.
25. Livi, S.; Duchet-Rumeau, J.; Gérard, J.F.; Pham, T.N. Polymers and ionic liquids: A successful wedding. *Macromol. Chem. Phys.* **2015**, *216*, 359–368. [CrossRef]
26. Ramesh, S.; Liew, C.W.; Ramesh, K. Evaluation and investigation on the effect of ionic liquid onto PMMA-PVC gel polymer blend electrolytes. *J. Non. Cryst. Solids.* **2011**, *357*, 2132–2138. [CrossRef]
27. Shamsuri, A.A.; Daik, R. Applications of ionic liquids and their mixtures for preparation of advanced polymer blends and composites: A. short review. *Rev. Adv. Mater. Sci.* **2015**, *40*, 45–59.
28. Plechkova, N.V.; Seddon, K.R. Applications of ionic liquids in the chemical industry. *Chem Soc Rev.* **2008**, *37*, 123–150. [CrossRef] [PubMed]
29. Lins, L.C.; Livi, S.; Duchet-Rumeau, J.; Gérard, J.F. Phosphonium ionic liquids as new compatibilizing agents of biopolymer blends composed of poly (butylene-adipate-co-terephtalate)/poly (lactic acid) (PBAT/PLA). *RSC Adv.* **2015**, *5*, 59082–59092. [CrossRef]
30. Stefanescu, C.; Daly, W.H.; Negulescu, I.I. Biocomposite films prepared from ionic liquid solutions of chitosan and cellulose. *Carbohydr. Polym.* **2012**, *87*, 435–443. [CrossRef]
31. Yousfi, M.; Livi, S.; Duchet-Rumeau, J. Ionic liquids: A new way for the compatibilization of thermoplastic blends. *Chem. Eng. J.* **2014**, *255*, 513–524. [CrossRef]
32. Livi, S.; Bugatti, V.; Marechal, M.; Soares, B.G.; Duchet-Rumeau, J.; Barra, G.M.O.; Gérard, J.F. Ionic Liquids-Lignin combination: An innovative way to improve mechanical behaviour and water vapour permeability of eco-designed biodegradable polymer blends. *RSC Adv.* **2015**, *5*, 1989–1998. [CrossRef]
33. Yang, J.; Pruvost, S.; Livi, S.; Duchet-Rumeau, J. Understanding of Versatile and Tunable Nanostructuration of Ionic Liquids on Fluorinated Copolymer. *Macromolecules* **2015**, *48*, 4581–4590. [CrossRef]
34. Lins, L.C.; Livi, S.; Maréchal, M.; Duchet-Rumeau, J.; Gerard, J.F. Structural dependence of cations and anions to building the polar phase of PVDF. *Eur. Pol. J.* **2018**, *107*, 236–248. [CrossRef]
35. Nguyen, T.K.L.; Livi, S.; Soares, B.G.; Pruvost, S.; Duchet-Rumeau, J.; Gérard, J.F. Ionic liquids: A New Route for the Design of Epoxy Networks. *ACS Sustain. Chem. Eng.* **2006**, *4*, 481–490. [CrossRef]

36. Soares, B.G.; Riany, N.; Silva, A.A.; Barra, G.M.O.; Livi, S. Dual-role of phosphonium–based ionic liquid in epoxy/MWCNT systems: Electric, rheological behavior and electromagnetic interference shielding effectiveness. *Eur. Polym. J.* **2016**, *84*, 77–88. [CrossRef]
37. Rahman, M.; Brazel, C.S. Ionic liquids: New generation stable plasticizers for poly (vinyl chloride). *Polym. Degrad. Stab.* **2006**, *91*, 3371–3382. [CrossRef]
38. Sankri, A.; Arhaliass, A.; Dez, I.; Gaumont, A.C.; Grohens, Y.; Lourdin, D.; Pillin, I.; Rolland-Sabaté, A.; Leroy, E. Thermoplastic starch plasticized by an ionic liquid. *Carbohydr. Polym.* **2010**, *82*, 256–263. [CrossRef]
39. Lins, L.C.; Bugatti, V.; Livi, S.; Gorrasi, G. Phosphonium ionic liquid as interfacial agent of layered double hydroxide: Application to a pectin matrix. *Carbohydr. Polym.* **2018**, *182*, 142–148. [CrossRef] [PubMed]
40. Lins, L.C.; Bugatti, V.; Livi, S.; Gorrasi, G. Ionic Liquid as Surfactant Agent of Hydrotalcite: Influence on the Final Properties of Polycaprolactone Matrix. *Polymers* **2018**, *10*, 44. [CrossRef] [PubMed]
41. Dias, A.M.A.; Marceneiro, S.; Braga, M.E.M.; Coelho, J.F.J.; Ferreira, A.G.M.; Simões, P.N.; Veiga, H.I.M.; Tomé, L.C.; Marrucho, I.M.; Esperança, J.M.S.S.; et al. Phosphonium-based ionic liquids as modifiers for biomedical grade poly (vinyl chloride). *Acta Biomater.* **2012**, *8*, 1366–1379. [CrossRef] [PubMed]
42. Chen, B.K.; Wu, T.Y.; Chang, Y.M.; Chen, A.F. Ductile polylactic acid prepared with ionic liquids. *Chem. Eng. J.* **2013**, *215*, 886–893. [CrossRef]
43. Scott, M.P.; Brazel, C.S.; Benton, M.G.; Mays, J.W.; Holbrey, D.; Rogers, R.D. Application of ionic liquids as plasticizers for poly (methyl methacrylate). *Chem. Commun.* **2002**, *13*, 1370–1371. [CrossRef] [PubMed]
44. Scott, M.P.; Rahman, M.; Brazel, C.S. Application of ionic liquids as low-volatility plasticizers for PMMA. *Eur. Polym. J.* **2003**, *39*, 1947–1953. [CrossRef]
45. Vyazovkin, S.; Burnham, A.K.; Criado, J.M.; Pérez-Maqueda, L.A.; Popescu, C.; Sbirrazzuoli, N. ICTAC Kinetics Committee recommendations for performing kinetic computations on thermal analysis data. *Thermochim. Acta* **2011**, *520*, 1–19. [CrossRef]
46. Ozawa, T. Estimation of activation energy by isoconversion methods. *Thermochim. Acta* **1992**, *203*, 159–165. [CrossRef]
47. Owens, D.K.; Wendt, R.C. Estimation of the surface free energy of polymers. *J. Appl. Polym. Sci.* **1969**, *13*, 1741–1747. [CrossRef]
48. Livi, S.; Gérard, J.F.; Duchet-Rumeau, J. Ionic liquids: Structuration agents in a fluorinated matrix. *Chem. Commun.* **2011**, *47*, 3589–3591. [CrossRef] [PubMed]
49. Kato, H.; Miki, K.; Mukai, T.; Nishikawa, K.; Koga, Y. Hydrophobicity/hydrophilicity of 1-butyl-2, 3-dimethyl and 1-ethyl-3-methylimodazolium ions: toward characterization of room temperature ionic liquids. *J. Phys. Chem. B* **2009**, *44*, 14754–14760. [CrossRef] [PubMed]
50. Vitucci, F.M.; Trequattrini, F.; Palumbo, O.; Brubach, J.B.; Roy, P.; Paolone, A. Infrared spectra of bis(trifluoromethanesulfonyl)imide based ionic liquids: Experiments and DFT simulations. *Vib. Spectrosc.* **2014**, *74*, 81–87. [CrossRef]
51. Herstedt, M.; Smirnov, M.; Johansson, P.; Chami, M.; Grondin, J.; Servant, L.; Lassègues, J.C. Spectroscopic characterization of the conformational states of the bis(trifluoromethanesulfonyl)imide anion (TFSI). *J. Raman Spectrosc.* **2005**, *36*, 762–770. [CrossRef]
52. Ueno, K.; Fukai, T.; Nagatsuka, T.; Yasuda, T.; Watanabe, M. Solubility of poly (methyl methacrylate) in ionic liquids in relation to solvent parameters. *Langmuir* **2014**, *30*, 3228–3235. [CrossRef] [PubMed]
53. Batista, M.L.S.; Neves, C.M.S.S.; Carvalho, P.J.; Gani, R.; Coutinho, J.A.P. Chameleonic behavior of ionic liquids and its impact on the estimation of solubility parameters. *J. Phys. Chem. B* **2011**, *115*, 12879–12888. [CrossRef] [PubMed]
54. Vitucci, F.M.; Trequattrini, F.; Palumbo, O.; Brubach, J.B.; Roy, P.; Navarra, M.A.; Panero, S.; Paolone, A. Stabilization of Different Conformers of Bis(trifluoromethanesulfonyl)imide Anion in Ammonium-Based Ionic Liquids at Low Temperatures. *J. Phys. Chem. A* **2014**, *118*, 8758–8764. [CrossRef]
55. Livi, S.; Gérard, J.F.; Duchet-Rumeau, J. Ionic Liquids as Polymer Additives. In *Applications of Ionic Liquids in Polymer Science Technology*; Mecerreyes, D., Ed.; Springer: Berlin/Heidelberg, Germany, 2015; pp. 1–21.
56. Maton, C.; de Vos, N.; Stevens, C.V. Ionic liquid thermal stabilities: Decomposition mechanisms and analysis tools. *Chem. Soc. Rev.* **2016**, *42*, 5963–5977. [CrossRef] [PubMed]
57. Huddleston, J.G.; Visser, A.E.; Reichert, W.M.; Willauer, H.D.; Broker, G.A.; Rogers, R.D. Characterization and comparison of hydrophilic and hydrophobic room temperature ionic liquids incorporating the imidazolium cation. *Green Chem.* **2001**, *3*, 156–164. [CrossRef]

58. Chan, B.K.M.; Chang, N.H.; Grimmett, M.R. The Synthesis and Thermolysis of Imidazole Quaternary Salts. *Aust. J. Chem.* **1977**, *30*, 2005–2013. [CrossRef]
59. Gordon, J.E. Fused Organic Salts. III. 1a Chemical Stability of Molten Tetra-n-alkylammonium Salts. Medium Effects on Thermal R4N+X- Decomposition. RBr + I- = RI + Br- Equilibrium Constant in Fused Salt Medium. *J. Org. Chem.* **1965**, *30*, 2760–2763. [CrossRef]
60. Kamavaram, V.; Reddy, R.G. Thermal stabilities of di-alkylimidazolium chloride ionic liquids. *Int. J. Therm. Sci.* **2008**, *47*, 773–777. [CrossRef]
61. Awad, W.H.; Gilman, J.W.; Nyden, M.; Harris, R.H.; Sutto, T.E.; Callahan, J.; Trulove, P.C.; DeLong, H.C.; Fox, D.M. Thermal degradation studies of alkyl-imidazolium salts and their application in nanocomposites. *Thermochim. Acta* **2004**, *409*, 3–11. [CrossRef]
62. Ngo, H.L.; LeCompte, K.; Hargens, L.; McEwen, A.B. Thermal properties of imidazolium ionic liquids. *Thermochim. Acta* **2000**, *357*, 97–102. [CrossRef]
63. Baranyai, K.J.; Deacon, G.B.; MacFarlane, D.R.; Pringle, J.M.; Scott, J.L. Thermal degradation of ionic liquids at elevated temperatures. *Aust. J. Chem.* **2004**, *57*, 145–147. [CrossRef]
64. Begg, C.G.; Grimmett, M.R.; Wethey, P.D. The thermally induced rearrangement of 1-substituted imidazoles. *Aust. J. Chem.* **1973**, *26*, 2435–2446. [CrossRef]
65. Lethesh, K.C.; Dehaen, W.; Binnemans, K. Base stable quaternary ammonium ionic liquids. *RSC Adv.* **2014**, *4*, 4472–4477. [CrossRef]
66. Kilaru, P.; Baker, G.A.; Scovazzo, P. Density and surface tension measurements of imidazolium-, quaternary phosphonium-, and ammonium-based room-temperature ionic liquids: Data and correlations. *J. Chem. Eng. Data* **2007**, *52*, 2306–2314. [CrossRef]
67. Freire, M.G.; Carvalho, P.J.; Fernandes, A.M.; Marrucho, I.M.; Queimada, A.J.; Coutinho, J.A.P. Surface tensions of imidazolium based ionic liquids: Anion, cation, temperature and water effect. *J. Colloid Interface Sci.* **2007**, *314*, 621–630. [CrossRef] [PubMed]
68. Santos, L.M.N.B.F.; Lopes, J.N.C.; Esperança, J.M.S.S.; Gomes, L.R.; Marrucho, I.M.; Rebelo, L.P.N. Ionic Liquids: First Direct Determination of their Cohesive Energy. *J. Am. Chem. Soc.* **2007**, *129*, 284–285. [CrossRef] [PubMed]
69. Kato, H.; Nishikawa, K.; Koga, Y. Relative hydrophobicity and hydrophilicity of some "Ionic liquid" Anions determined by the 1-propanol probing methodology: A differential thermodynamic approach. *J. Phys. Chem. B* **2008**, *112*, 2655–2660. [CrossRef] [PubMed]
70. Zhao, L.; Li, Y.; Cao, X.; You, J.; Dong, W. Multifunctional role of an ionic liquid in melt-blended poly (methyl methacrylate)/multi-walled carbon nanotube nanocomposites. *Nanotechnology* **2012**, *23*, 255702. [CrossRef] [PubMed]
71. Wojciechowska, P. The Effect of Concentration and Type of Plasticizer on the Mechanical Properties of Cellulose Acetate Butyrate Organic-Inorganic Hybrids. In *Recent Advances in Plasticizers*; Luqman, M., Ed.; InTech: Vienna, Austria, 2012; pp. 141–164.
72. Luo, W.; Liu, W. Incubation time to crazing in stressed poly (methyl methacrylate). *Polym. Test.* **2007**, *26*, 413–418. [CrossRef]
73. Andrews, E.H.; Bevan, L. Mechanics and mechanism of environmental crazing in a polymeric glass. *Polymer* **1972**, *13*, 337–346. [CrossRef]
74. Bucknall, C.B. Role of surface chain mobility in crazing. *Polymer* **2012**, *53*, 4778–4786. [CrossRef]
75. Scheirs, J. *Compositional and Failure Analysis of Polymers: A Practical Approach*, 1st ed.; John Wiley & Sons: New York, NY, USA, 2000.
76. Plummer, C.J.G.; Donald, A.M. Crazing mechanisms and craze healing in glassy polymers. *J. Mater. Sci.* **1989**, *24*, 1399–1405. [CrossRef]
77. Mahajan, D.K.; Hartmaier, A. Mechanisms of crazing in glassy polymers revealed by molecular dynamics simulations. *Phys. Rev. E* **2012**, *86*, 021802. [CrossRef] [PubMed]

Article

Investigation on the Crystallization Behaviors of Polyoxymethylene with a Small Amount of Ionic Liquid

Qi Jiao [1,2,3], Qin Chen [4], Lian Wang [4,*], Hualin Chen [1,2] and Yongjin Li [4]

1 Chengdu Institute of Organic Chemistry, Chinese Academy of Sciences, Chengdu 610041, China; jiaoqisncc@sina.com (Q.J.); aofly@163.com (H.C.)
2 University of Chinese Academy of Sciences, Beijing 100049, China
3 Coal Chemical Industry Technology Research Institute, Shenhua Ningxia Coal Group Co. Ltd., Yinchuan 750411, China
4 College of Materials, Chemistry and Chemical Engineering, Hangzhou Normal University, No. 2318 Yuhangtang Rd., Hangzhou 311121, China; 15757116907@163.com (Q.C.); yongjin-li@hznu.edu.cn (Y.L.)
* Correspondence: wanglian@hznu.edu.cn; Tel.: +86-571-2886-6579

Received: 20 December 2018; Accepted: 4 February 2019; Published: 5 February 2019

Abstract: Polyoxymethylene (POM) blends with excellent stiffness–toughness balance are successfully developed using Tributyl(octyl)phosphonium bis(trifloromethanesulfonyl) imide (TBOP-TFSI), one type of room-temperature ionic liquid, as the nucleating agent. Crystallization behaviors of POM blends have been studied by differential scanning calorimetry (DSC) and polarized light microscopy (PLM). The incorporation of TBOP-TFSI induces the crystal nucleation and fine crystal grain of POM, and also a much shorter hemi-crystalline time with only 0.5 wt% addition. The nucleation effect of ionic liquid leads to considerable improvement in the impact strength of POM blends while not sacrificing its tensile strength. Moreover, antistatic properties with a long-time stable performance are achieved by TBOP-TFSI addition as the electrical resistance reaches 10^{11} Ω/sq.

Keywords: polyoxymethylene; ionic liquid; crystallization behavior; nucleation

1. Introduction

Polyoxymethylene (POM) is one of the most important engineering plastics and is broadly used in many areas, such as automobiles, the coal industry, and also in architecture materials [1–6]. POM is a kind of weakly polar linear crystalline polymer, and its molecular structure is: $[CH_2O]_n$. Due to its highly regularly structured molecular chain, POM can easily crystallize to form spherulites, and exhibits high crystallinity. The main reason for its excellent performance is that POM has the high density and crystallinity. For example, POM has a high strength and flexural modulus, excellent chemical and creep resistance, and excellent dimensional stability, wear and friction properties. Though high crystallinity of POM brings excellent mechanical properties, the large spherulites of POM also lead to low impact toughness and gap-sensitivity. Moreover, POM has a high surface resistivity of about 10^{14} to 10^{17} Ω/sq. Such high resistivity leads to easily accumulated electrostatic charges on the surface of POM. Such features limit the application of POM in industry and daily life [4–10]. Improving its crystallization performance can enhance the properties of POM and expand the application in industry.

Many scientists have investigated the effects of nucleating agents on the crystallization performance of POM [6,11–19]. Zhao [19] showed that multi-walled carbon nanotubes (MWCNTs) had an excellent nucleation effect on POM, which lead to an increase in crystallization temperature and crystallization rate. Hu [20] showed the isothermal and non-isothermal crystallization behavior and morphology of POM blended with a small amount of polyamide (PA) as the nucleating agent.

They illustrated that the addition of PA reduced the spherulite size and improved the crystallization growth rate and the crystallinity of POM due to the nucleation effect of PA on POM.

Recently, ionic liquids, have been proposed as lubricants [21], plasticizers [22], and surfactants [23] in many polymer blends as a new type of modification agent due to their low melting point, low toxicity, and high chemical and physical stability. Ionic liquids have also been reported as nucleating agents for a few semicrystalline polymers. Liu [24] reported one pyrrolidinium ionic liquid served as the nucleating agents for PET and shortened induction time and enhanced crystallization rate. Previously, it is reported by our group that a kind of ILs, 1-butyl-3-methylimidazolium hexafluorophosphate [BMIM][PF6], can increase the γ-phase content of poly(vinylidene fluoride) (PVDF) by simply mix blending, which indicates [BMIM][PF6] played a role as a γ-phase nucleating agent in these blends [25]. However, literatures on the nucleation effect of ionic liquids on crystallization behavior for other semicrystalline polymers are quite limited.

Moreover, ionic liquids have been utilized as antistatic agents due to their excellent ionic conductivity. Combining conductive agents into insulative polymers is of some importance to prevent electrostatic-charge accumulation on the surface of neat polymers, especially for those applied in electronic devices [26,27]. Pernak [28] et al. first incorporated ionic liquid into polyethylene matrix to investigate antistatic ability related to the chemical structure of ionic liquid. Since then, ionic liquids have been reported as antistatic agents for various polymers, such as polypropylene [12,29], poly(vinylidene flouride) [25,30], polystyrene [31], and also biodegradable poly(L-lactic acid) (PLLA) [32]. It is believed compatibility and interaction between ionic liquid and matrix are of crucial importance when designing antistatic composites. Ionic liquids with moderate compatibility or intermolecular interactions should be considered as suitable modification agents.

In this work, we used Tributyl(octyl)phosphonium bis(trifloro methanesulfonyl) imide (TBOP-TFSI, short for convenience as TBOP-TFSI), one type of room-temperature ionic liquid (RTIL), as a nucleating agent and antistatic agent, which showed a partial compatibility and a weak ion–molecular interaction with the matrix to obtain multi-functional POM materials by simple mixing. It was found that small amounts of TBOP-TFSI could facilitate the nucleation of POM, and decrease the sizes of POM spherulites, which resulted in the improvement in the toughness of POM. More importantly, the tensile strength of composites did not decrease. Moreover, the POM composite achieved antistatic effects by adding only 0.5 wt% TBOP-TFSI. In a word, we fabricated antistatic POM composites with simultaneously excellent stiffness–toughness balance.

2. Materials and Methods

2.1. Materials

POM (MC 90) used was commercially available from Shenhua Co., Ltd. (Yinchuan China). TBOP-TFSI was supplied by Solvay Co., Ltd (Brussels, Belgium). All of the samples were used as received.

2.2. Preparation of POM/TBOP-TFSI Blends

All materials were vacuum-dried at 80 °C for 12 h before use. The composites were prepared by extruder with single screw at 170 to 200 °C with a rotation speed of 50 rpm after simply mixing physically. After Extrusion, some samples were hot-pressed at 200 °C under a 10 MPa pressure for 3 min to a 300 μm film, followed by cool-pressing for 2 min at room temperature. The obtained sheets were used for the following characterization. The specimens for tensile tests and notched impact test were prepared by injection molding. The IL loadings in this work were calculated based on only the amount of matrix POM by weight.

2.3. Characterization

2.3.1. Antistatic Properties

Electrical conductivity was tested by an ultrahigh resistivity meter, using a piece of URS probe electrode (MCPHT450) at 100 V with a 300 μm-thickness sample.

2.3.2. Mechanical Properties

Tensile test were measured by Instron universal material testing system (model 5966) at 25 °C. The tensile rate was 5 mm/min. The notched impact test was performed according to the GB/T16420-1996 standard. Samples for tensile test and notched impact test were prepared by injection molding (Haake MINIJET PRO).

2.3.3. Scanning Electron Microscopy

The microstructure of cross-fractured surface of samples was obtained using field emission scanning electron microscopy (FESEM, SEM-JSM 6700). The acceleration voltage was 3 kV and the fractured surface was coated with a thin layer of gold before the SEM observation.

2.3.4. Fourier Transfrom Infrared (FTIR)

The Fourier transfrom infrared (FTIR) measurements were carried out in transmittance mode on grinding samples by FTIR spectroscopy (FTIR, Bruker Tensor). The FTIR spectra were recorded at a resolution of 2 cm^{-1}, and 64 scans from 4000 to 400 cm^{-1} were averaged. The samples were prepared by spin-coating after dissolving neat POM and the blends in hexafluoroisopropanol.

2.3.5. Differential Scanning Calorimetry

The differential scanning calorimetry (DSC) measurements were carried out by a differential scanning calorimetry (DSC, TA-Q2000). Some samples were first heated to 200 °C and held for 5 min to eliminate previous thermal history, then cooled to 20 °C. Both the cooling and heating rates were 10 °C/min. The others were first heated to 200 °C like the previous one, then cooled to 152 °C as fast as possible for isothermal crystallization process. The experiments were carried out under a continuous high purity nitrogen atmosphere.

2.3.6. Polarized Light Microscopy

The morphologies of the POM spherulites were observed by polarizing light microscopy (PLM, Olympus BX51) equipped with a digital camera, using a Linkham LTS 350 hot stage to control the temperature at 152 °C. All samples were spin-coated onto clean glass sheets, and the thickness of samples on the glass sheet was about 10 mm.

3. Results

3.1. Morphologies of POM/TBOP-TFSI Blends and the Interactions between POM and TBOP-TFSI

Figure 1 shows SEM images of fracture surface morphologies of POM/TBOP-TFSI thin films with indicated TBOP-TFSI loadings. No visible phase was observed for the blend containing 0.5 wt% ionic liquid before or after methanol extraction, suggesting a homogeneous distribution of TBOP-TFSI within POM matrix. However, several domains of less than 100 nanometers were observed for blending with 1 wt% ionic liquid loading, indicating a phase-separated structure. Increasing TBOP-TFSI loading induced the aggregation of ionic liquid and domain formation in the matrix. A typical sea-island structure was observed at 3 wt% TBOP-TFSI loading, of which domain size of ionic liquid became over several hundred nanometers. Some anomalous domains could be seen for this sample after extracting ionic liquid by methanol at 98 °C for 24 h, implying the partial compatibility nature of POM and

TBOP-TFSI. It is believed that the dispersion of TBOP-TFSI within matrix was responsible for several properties as surface conductivity of binary blends. POM and TBOP-TFSI remained compatible at very low ionic liquid content. Consequently, an electrical conducting pathway could be formed due to the high mobility of ionic liquid. However, for blends containing more than 1 wt% of ionic liquid, typical sea-island structures were formed due to phase separation between ionic liquid and matrix.

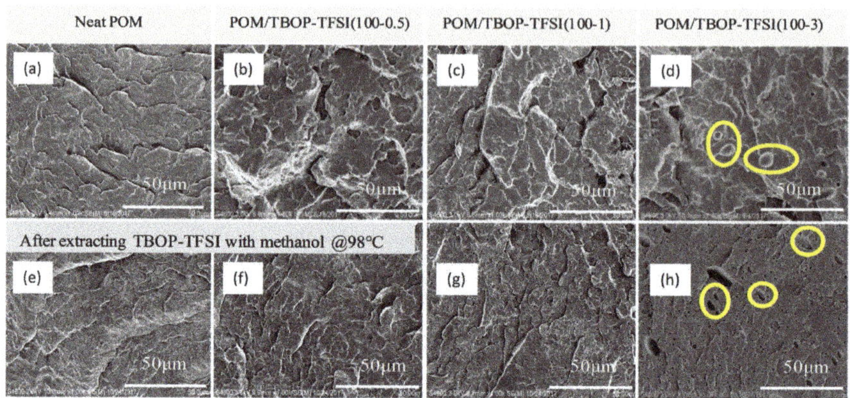

Figure 1. SEM images of fracture surface morphologies of polyoxymethylene (POM)/ tributyl(octyl)phosphonium bis(trifloromethanesulfonyl) imide (TBOP-TFSI) blends with indicated TBOP-TFSI loadings before (**a**–**d**) and after extracting with methanol (**e**–**h**).

We then utilized FTIR to observe the intermolecular interaction between POM and TBOP-TFSI. Figure 2a shows the molecular structure of TBOP-TFSI, Figure 2b shows the FTIR spectra of POM/TBOP thin films. It is found that the absorption bands appearing at 2922 cm^{-1} and 2790 cm^{-1} are due to the symmetric stretching vibration of C-H, and the absorption peak at 2979 cm^{-1} is assigned to the C-H asymmetric stretching vibration of POM [33]. The three peaks shift to lower wavenumbers for POM/TBOP-TFSI mixture samples, and the absorption peaks for POM/TBOP-TFSI samples are at 2975, 2917, and 2788 cm^{-1}, respectively, indicating the electrostatic interaction between TBOP-TFSI and POM. Because of the existence of this interaction, POM and TBOP-TFSI are partially miscible, consisting of the morphology of composites (as shown in Figure 1).

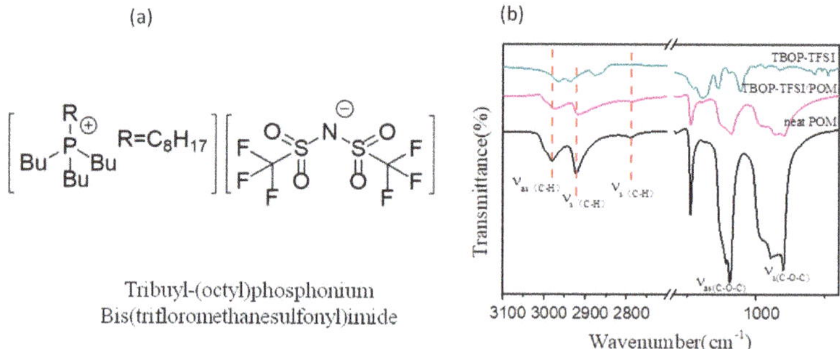

Figure 2. (**a**) Molecular structure of TBOP-TFSI; (**b**) Fourier transfrom infrared (FTIR) spectra of POM with TBOP-TFSI.

3.2. Crystallization Behaviors of POM/TBOP-TFSI Blends

The effects of TBOP-TFSI on the crystallization of POM have been investigated by isothermal and non-isothermal crystallization. It was found that TBOP-TFSI improves both the crystallization temperature and degree of crystallinity of POM as Table 1 shows, indicating that the incorporation of TBOP-TFSI accelerates the crystallization of POM. The crystallinity of POM in all samples was determined by Equation (1).

$$\chi_c = \Delta H_m / (W_m \times \Delta H_m^\circ) \times 100\% \quad (1)$$

where ΔH_m° is the theoretical melting enthalpy of POM with a value of 247 J/g [9] and W_m is the weight fraction of matrix in composites. Figure 3c,d reveal that the isothermal crystallization time for POM blending with TBOP-TFSI is shorter than that of neat POM. For the POM blended with 0.5 wt% TBOP-TFSI, the $t_{1/2}$ is about 2.4 min, while it takes neat POM for about 5.9 min at 152 °C. The isothermal crystallization behaviors of POM/TBOP-TFSI blends are consistent with the non-isothermal crystallization behaviors, as shown in Figure 3a,b. What is more, the value of (T_m−T_c) for POM becomes lower, which means the incorporation of TBOP-TFSI results in a fine crystal grain. From both isothermal and non-isothermal crystallization of POM/TBOP-TFSI thin films, 0.5 wt% addition of ionic liquid, showed a better nucleating ability than other content. We believed this was due to homegeneous dispersion of ionic liquid within the matrix as they were compatible with less than 0.5 wt% content, and also responsible for enhanced mechanical performance.

Table 1. Differential scanning calorimetry (DSC) results of polyoxymethylene (POM)/ tributyl(octyl)phosphonium bis(trifloromethanesulfonyl) imide (TBOP-TFSI) composites.

Samples	T_c (°C)	T_m (°C)	T_m−T_c (°C)	$t_{1/2}$ (min) at 152 °C	χ_c (%)
Neat POM	145.5	165.4	19.9	5.9	51.1
POM/TBOP-TFSI (100-0.25)	145.7	164.9	19.2	6.4	51.3
POM/TBOP-TFSI (100-0.5)	147.1	165.2	18.1	2.4	53.2
POM/TBOP-TFSI (100-1)	146.2	165.0	18.8	5.5	53.6
POM/TBOP-TFSI (100-3)	146.8	164.9	18.1	5.3	52.9
POM/TBOP-TFSI (100-5)	146.8	164.8	18.0	5.3	52.0

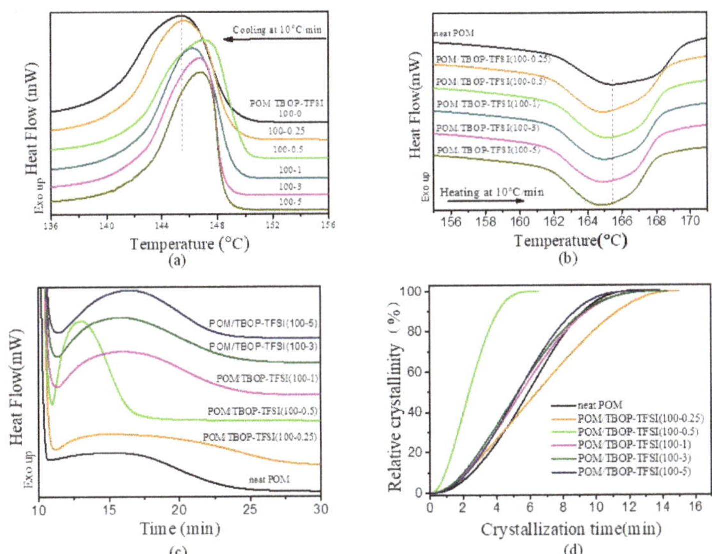

Figure 3. Differential scanning calorimetry (DSC) curves for POM/TBOP-TFSI composites: (**a**) first cooling and (**b**) second heating (after 5 min isothermal for erasing heat history); (**c**) isothermal crystallization process at 152 °C; (**d**) relative crystallinity as a function of crystallization time at 152 °C.

The nucleation effect of the ionic liquid can also be confirmed by optical polarize microscopy as in Figure 4. The sizes of POM spherulites became smaller with more loading of TBOP-TFSI (370 μm of neat POM spherulites decreased to 98 μm with 3 wt% of TBOP-TFSI loading). With higher crystal nucleation density and fine crystal grain, the mechanical properties of POM blends achieved a satisfactory enhancement.

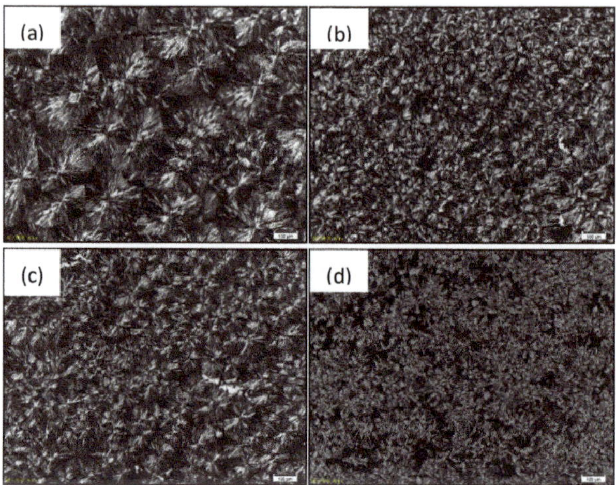

Figure 4. Polarized light microscopy (PLM) images of (**a**) neat POM; (**b**) POM/TBOP-TFSI (100-0.5); (**c**) POM/TBOP-TFSI (100-1); (**d**) POM/TBOP-TFSI (100-3) isothermally crystallized at 152 °C.

3.3. Mechanical Properties of POM/TBOP-TFSI Blends

The strain–stress curves of POM and POM/TBOP-TFSI specimens are shown in Figure 5 (Table 2). One can observe a significant enhancement of elongation at the break of POM with the addition of a small amount of TBOP-TFSI. At the same time, the tensile strength has a slight improvement. Furthermore, the impact strength of the blends also increases to 4.7 kJ/m^2 with 0.5 wt% TBOP-TFSI loading as compared with neat POM (4.25 kJ/m^2), as shown in the exact values of mechanical properties listed in Table 2. In general, the addition of a small amount TBOP-TFSI not only improves the ductility but also maintains the strength of POM matrix. We believe that the smaller sizes of spherulites induced by ionic liquid nucleation were responsible for improved mechanical performance, especially impact strength. However, the addition of more ionic liquid induced worse mechanical properties due to phase separation between TBOP-TFSI and matrix as showed in Figure 2.

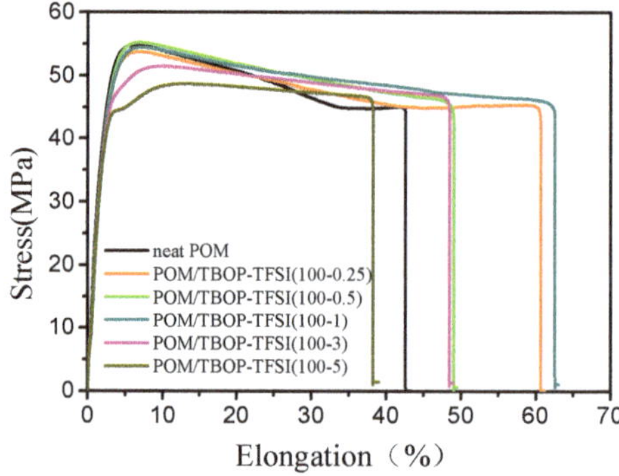

Figure 5. Stress-elongation curves of POM/TBOP-TFSI composites.

Table 2. Mechanical properties of POM/TBOP-TFSI composites.

Sample	Yeilding Strength (MPa)	Elongation at Break (%)	Impact Strength (kJ/m^2)
Neat POM	55.2 ± 0.8	40.2 ± 10	4.2
POM/TBOP-TFSI (100-0.25)	53.8 ± 0.7	56.1 ± 11	4.6
POM/TBOP-TFSI (100-0.5)	54.7 ± 0.5	47.0 ± 9	4.7
POM/TBOP-TFSI (100-1)	54.2 ± 0.8	63.3 ± 12	4.5
POM/TBOP-TFSI (100-3)	51.5 ± 0.5	42.0 ± 6	4.4
POM/TBOP-TFSI (100-5)	48.3 ± 1.0	39.2 ± 13	4.5

3.4. Antistatic Properties of POM/TBOP-TFSI Blends

The combination of ionic liquid usually results in the antistatic property for polymer blends. Figure 6 shows the electrical conductivity of POM/TBOP-TFSI films corresponding to TBOP-TFSI loading. It is illustrated that the electrical conductivity of POM/TBOP-TFSI blends decreases with the increasing of TBOP-TFSI content. Generally, the surface resistivity of antistatic material is less than 10^{12} Ω/sq, and neat POM is an insulating material with a surface resistivity higher than 10^{13} Ω/sq. It could be observed in Figure 6 that trace amounts of ionic liquid would be adequate for fabricating antistatic POM materials. The antistatic property can be achieved with only 0.5 wt% TBOP-TFSI addition which indicates the formation of a continuous conducting network within the matrix. The surface resistivity could be further reduced by increasing TBOP-TFSI loading. For blends containing 3 wt% of TBOP-TFSI, the surface resistivity is 3.78×10^9 Ω/sq, which was 2 orders of magnitude less than the 0.5 wt% one. Although electrical conductivity of POM blends could be improved by higher addition of TBOP-TFSI, an excess ionic liquid bleeding was observed for blends with more than 3 wt% TBOP-TFSI loading during mixing for their moderate compatibility with POM, which is unfavorable for mechanical property. Moreover, the values of surface resistivity of all blends stayed almost the same or even lower after storage for a few weeks. The stable antistatic property resulted from partial compatibility of TBOP-TFSI and POM matrix (Figure 1). Ionic liquids were found miscible with POM at very low content (0.5 wt%), leading to the formation of a conductive path due to homogeneous dispersion of TBOP-TFSI in the matrix. At relatively high loadings (1 wt%), although TBOP-TFSI started to aggregate, there were still no large domains observed. One can still obtain the reduced surface resistivity. This excellent long term antistatic performance is of crucial importance for practical applications as internal dust-free parts in electronic devices and industrial structural components that meet the requirement of both mechanical strength and dissipation of static electricity.

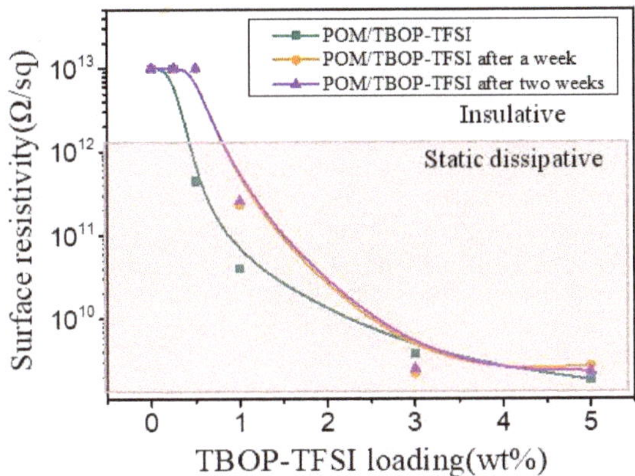

Figure 6. Surface resistivity of POM/TBOP-TFSI composites.

4. Conclusions

We investigated crystallization behavior of POM with a very small amount of ionic liquid in binary blends. In POM/TBOP-TFSI blends, ionic liquid acted as a nucleation agent for POM molecules, leading to a higher crystallization temperature and a shorter crystallization time. The size of POM spherulites decreased magnificently as nucleation density increased with ionic liquid addition. Due to the nucleation effect, enhanced mechanical properties could be achieved for POM/TBOP-TFSI binary blends. Furthermore, binary blends exhibited stable surface conductivity which might extend the practical application of POM.

Author Contributions: L.W. and Y.L. conceived the idea and designed experiments; Q.J. and Q.C. performed the experiments; H.C. and Y.L. supervised the work; Q.J. wrote the paper. All the authors contributed to the preparation of the manuscript.

Funding: This research was funded by Zhejiang Province Natural Science Foundation (LQ14B040004) and the National Natural Science Foundation of China (51403046).

Conflicts of Interest: The authors declare no conflict of interest.

References

1. Hama, H.; Tashiro, K. Structural changes in non-isothermal crystallization process of melt-cooled polyoxymethylene. [I] Detection of infrared bands characteristic of folded and extended chain crystal morphologies and extraction of a lamellar stacking model. *Polymer* **2003**, *44*, 3107–3116. [CrossRef]
2. Hama, H.; Tashiro, K. Structural changes in isothermal crystallization process of polyoxymethylene investigated by time-resolved FTIR, SAXS and WAXS measurements. *Polymer* **2003**, *44*, 6973–6988. [CrossRef]
3. Hasegawa, S.; Takeshita, H.; Yoshii, F.; Sasaki, T.; Makuuchi, K.; Nishimoto, S. Thermal degradation behavior of gamma-irradiated acetyloxy end-capped poly(oxymethylene). *Polymer* **2000**, *41*, 111–120. [CrossRef]
4. Jiang, Z.; Chen, Y.; Liu, Z. The morphology, crystallization and conductive performance of a polyoxymethylene/ carbon nanotube nanocomposite prepared under microinjection molding conditions. *J. Polym. Res.* **2014**, *21*, 451. [CrossRef]
5. Li, K.; Xiang, D.; Lei, X. Green and self-lubricating polyoxymethylene composites filled with low-density polyethylene and rice husk flour. *J. Appl. Polym. Sci.* **2010**, *108*, 2778–2786. [CrossRef]
6. Erol, B.; Ibrahim Unal, H.; Sari, B. Electrorheology and creep-recovery behavior of conducting polythiophene/poly(oxymethylene)-blend suspensions. *Chin. J. Polym. Sci.* **2012**, *30*, 16–25. [CrossRef]
7. Archodoulaki, V.M.; Lüftl, S.; Seidler, S. Thermal degradation behaviour of poly(oxymethylene): 1. Degradation and stabilizer consumption. *Polym. Degrad. Stab.* **2004**, *86*, 75–83. [CrossRef]
8. Lüftl, S.; Archodoulaki, V.M.; Seidler, S. Thermal-oxidative induced degradation behaviour of polyoxymethylene (POM) copolymer detected by TGA/MS. *Polym. Degrad. Stab.* **2006**, *91*, 464–471. [CrossRef]
9. Xu, W.; He, P. Isothermal crystallization behavior of polyoxymethylene with and without nucleating agents. *J. Appl. Polym. Sci.* **2001**, *80*, 304–310. [CrossRef]
10. Bao, H.D.; Guo, Z.X.; Yu, J. Effect of electrically inert particulate filler on electrical resistivity of polymer/multi-walled carbon nanotube composites. *Polymer* **2008**, *49*, 3826–3831. [CrossRef]
11. Cheng, Z.; Wang, Q. Morphology control of polyoxy-methylene/thermoplastic polyurethane blends by adjusting their viscosity ratio. *Polym. Int.* **2010**, *55*, 1075–1080. [CrossRef]
12. Ding, Y.; Tang, H.; Zhang, X.; Wu, S.; Xiong, R. Antistatic ability of 1-n-tetradecyl-3-methylimidazolium bromide and its effects on the structure and properties of polypropylene. *Eur. Polym. J.* **2008**, *44*, 1247–1251. [CrossRef]
13. Mehrabzadeh, M.; Rezaie, D. Impact modification of polyacetal by thermoplastic elastomer polyurethane. *J. Appl. Polym. Sci.* **2010**, *84*, 2573–2582. [CrossRef]
14. Pecorini, T.J.; Hertzberg, R.W.; Manson, J.A. Structure-property relations in an injection-moulded, rubber-toughened, semicrystalline polyoxymethylene. *J. Mater. Sci.* **1990**, *25*, 3385–3395. [CrossRef]
15. Pielichowski, K.; Leszczynska, A. Structure-Property Relationships in Polyoxymethylene/Thermoplastic Polyurethane Elastomer Blends: Journal of Polymer Engineering. *J. Polym. Eng.* **2011**, *25*, 160–165.

16. Sumita, M.; Sakata, K.; Asai, S.; Miyasaka, K.; Nakagawa, H. Dispersion of fillers and the electrical conductivity of polymer blends filled with carbon black. *Polym. Bull.* **1991**, *25*, 265–271. [CrossRef]
17. Tang, G.; Shao, C.; Hu, X.; Tang, T. The mechanical properties of carbon fiber-reinforced polyoxymethylene composites filled with silaned nano-SiO_2. *J. Thermoplast. Compos. Mater.* **2014**, *29*, 1020–1029. [CrossRef]
18. Zhao, X.; Zhen, L.; Jian, L.; Fei, F.Y. The effect of CF and nano-SiO2 modification on the flexural and tribological properties of POM composites. *J. Thermoplast. Compos. Mater.* **2012**, *27*, 287–296.
19. Zhao, X.; Ye, L. Structure and mechanical properties of polyoxymethylene/multi-walled carbon nanotube composites. *Compos. Part B* **2011**, *42*, 926–933. [CrossRef]
20. Hu, Y.; Ye, L. Nucleation effect of polyamide on polyoxymethylene. *Polym. Eng. Sci.* **2010**, *45*, 1174–1179. [CrossRef]
21. Sanes, J.; Carrion, F.; Bermudez, M.D.; Martineznicolas, G. Ionic liquids as lubricants of polystyrene and polyamide 6-steel contacts. Preparation and properties of new polymer-ionic liquid dispersions. *Tribol. Lett.* **2006**, *21*, 121–133. [CrossRef]
22. Li, X.; Qin, A.; Zhao, X.; Ma, B.; He, C. The plasticization mechanism of polyacrylonitrile/1-butyl-3-methylimidazolium chloride system. *Polymer* **2014**, *55*, 5773–5780. [CrossRef]
23. Lins, L.C.; Bugatti, V.; Livi, S.; Gorrasi, G. Ionic Liquid as Surfactant Agent of Hydrotalcite: Influence on the Final Properties of Polycaprolactone Matrix. *Polymer* **2018**, *10*, 44. [CrossRef]
24. Dou, J.; Liu, Z. Crystallization behavior of poly(ethylene terephthalate)/pyrrolidinium ionic liquid. *Polym. Int.* **2013**, *62*, 1698–1710. [CrossRef]
25. Xing, C.; Zhao, M.; Zhao, L.; You, J.; Cao, X.; Li, Y. Ionic liquid modified poly(vinylidene fluoride): crystalline structures, miscibility, and physical properties. *Polym. Chem.* **2013**, *4*, 5726–5734. [CrossRef]
26. Tsurumaki, A.; Tajima, S.; Iwata, T.; Scrosati, B.; Ohno, H. Antistatic effects of ionic liquids for polyether-based polyurethanes. *Electrochim. Acta* **2015**, *175*, 13–17. [CrossRef]
27. Iwata, T.; Tsurumaki, A.; Tajima, S.; Ohno, H. Fixation of ionic liquids into polyether-based polyurethane films to maintain long-term antistatic properties. *Polymer* **2014**, *55*, 2501–2504. [CrossRef]
28. Pernak, J.; Sobaszkiewicz, K.; Foksowicz-Flaczyk, J. Ionic liquids with symmetrical dialkoxymethyl-substituted imidazolium cations. *Chemistry* **2004**, *10*, 3479–3485. [CrossRef]
29. Wang, X.; Liu, L.; Tan, J. Preparation of an ionic-liquid antistatic/photostabilization additive and its effects on polypropylene. *J. Vinyl Addit. Technol.* **2010**, *16*, 58–63. [CrossRef]
30. Mejri, R.; Dias, J.C.; Lopes, A.C.; Hentati, S.B.; Silva, M.M.; Botelho, G.; De Ferro, A.M.; Esperanca, J.M.S.S.; Maceiras, A.; Laza, J.M. Effect of ionic liquid anion and cation on the physico-chemical properties of poly(vinylidene fluoride)/ionic liquid blends. *Eur. Polym. J.* **2015**, *71*, 304–313. [CrossRef]
31. Lu, X.; Zhou, J.; Zhao, Y.; Qiu, Y.; Li, J. Room Temperature Ionic Liquid Based Polystyrene Nanofibers with Superhydrophobicity and Conductivity Produced by Electrospinning. *Chem. Mater.* **2008**, *20*, 3420–3424.
32. Wei, T.; Pang, S.; Xu, N.; Pan, L.; Zhang, Z.; Xu, R.; Ma, N.; Lin, Q. Crystallization behavior and isothermal crystallization kinetics of PLLA blended with ionic liquid, 1-butyl-3-methylimidazolium dibutylphosphate. *J. Appl. Polym. Sci.* **2015**, *132*, 41308–41319. [CrossRef]
33. Yu, N.; He, L.; Ren, Y.; Xu, Q. High-crystallization polyoxymethylene modification on carbon nanotubes with assistance of supercritical carbon dioxide: Molecular interactions and their thermal stability. *Polymer* **2011**, *52*, 472–480. [CrossRef]

Article

Gelled Electrolyte Containing Phosphonium Ionic Liquids for Lithium-Ion Batteries

Mélody Leclère [1,2], Laurent Bernard [2], Sébastien Livi [1], Michel Bardet [3], Armel Guillermo [2], Lionel Picard [2] and Jannick Duchet-Rumeau [1,*]

1 Université de Lyon, F-69003, Lyon, France, INSA Lyon, CNRS, UMR 5223, Ingénierie des Matériaux Polymères, F-69621 Villeurbanne, France; melody.leclere@insa-lyon.fr (M.L.); sebastien.livi@insa-lyon.fr (S.L.)

2 Université Grenoble Alpes, CEA-Grenoble, LITEN/DEHT/SCGE/LGI, 17, F-38000 Grenoble, France; Laurent.BERNARD3@cea.fr (L.B.); armel.guillermo@free.fr (A.G.); lionel.picard@cea.fr (L.P.)

3 Université Grenoble Alpes, CEA-Grenoble, INAC/SCIB/LRM, F-38000 Grenoble, France; michel.bardet@cea.fr

* Correspondence: jannick.duchet@insa-lyon.fr; Tel.: +33-472-438-291

Received: 16 May 2018; Accepted: 12 June 2018; Published: 14 June 2018

Abstract: In this work, new gelled electrolytes were prepared based on a mixture containing phosphonium ionic liquid (IL) composed of trihexyl(tetradecyl)phosphonium cation combined with bis(trifluoromethane)sulfonimide [TFSI] counter anions and lithium salt, confined in a host network made from an epoxy prepolymer and amine hardener. We have demonstrated that the addition of electrolyte plays a key role on the kinetics of polymerization but also on the final properties of epoxy networks, especially thermal, thermo-mechanical, transport, and electrochemical properties. Thus, polymer electrolytes with excellent thermal stability (>300 °C) combined with good thermo-mechanical properties have been prepared. In addition, an ionic conductivity of 0.13 $Ms{\cdot}cm^{-1}$ at 100 °C was reached. Its electrochemical stability was 3.95 V vs. Li^0/Li^+ and the assembled cell consisting in Li | $LiFePO_4$ exhibited stable cycle properties even after 30 cycles. These results highlight a promising gelled electrolyte for future lithium ion batteries.

Keywords: ionic liquids; thermosets; Lithium salts; electrolytes

1. Introduction

These last years, the development of storage energy system such as lithium ion batteries has witnessed a real boom in academic and industrial research [1,2]. In fact, rechargeable batteries with high energy/power density, cycling stability, and safety were developed [3,4]. However, many challenges must be overcome, such as the development of safe electrolytes [5]. In this case, gelled electrolytes represent a promising way to prevent leakage of liquid electrolyte when the electrolytes are used in electrochemical devices. Indeed, gelled electrolytes confine the liquid electrolyte in a polymer matrix, which plays the role of host. Thus, many works were reported and were investigated from the confinement of a conventional electrolyte (organic solvent with lithium salt) and extended to the use of ionic liquids (ILs) as a solvent [6,7]. Recently, ionic liquids (ILs) have appeared as promising candidates to replace flammable organic solvents thanks to their excellent thermal and chemical stability, their good ionic conductivity, their low vapor pressure, and their endless cation/anion combinations [8–11]. Moreover, some ILs have presented a high compatibility with linear polymer [12–14], such as polyethylene oxide (PEO), polyvinylidene fluoride-*co*-hexafluoropropylene (PVDF-HFP), poly(ionic liquids) PILs (polydiallylmethylammonium), and thermosets such as epoxy networks [15–19]. The use of a crosslinked polymer as a host polymer is an interesting route to develop polymer electrolytes with good mechanical performances despite a large quantity of introduced ionic liquids. Moreover, the main advantage of these gelled electrolytes is that no solvent is required during their processing. For example, Stepniak et al. have prepared a gelled electrolyte based on *N*-methyl-*N*-propylpiperidinium [MPPip] trifluroromethanesulfonimide [TFSI] obtained by photopolymerization of bisphenol A diacryalate [18]. These authors have highlighted the significant mechanical strength of the electrolyte even with 80 wt % of confined electrolyte and an ionic conductivity of 0.63 mS·cm^{-1}. In addition, a good electrochemical performance in Li | LFP battery at 25 °C with moderate rate C/10 getting a specific capacity superior to 150 m Ah·g^{-1} was demonstrated. Other authors such as Sotta et al. have developed a gelled electrolyte based on epoxy networks (DGEBA/polyetheramine) containing 1-Butyl-3-methylimidazolium bis(trifluoromethyl-sulfonyl)imide [BMIM] [TFSI] and have obtained excellent transport properties with an ionic conductivity of approximately 0.4 mS·cm^{-1} at 20 °C [20]. However, the electrochemical stability window of imidazolium-based ILs being limited to 1.5–5 V vs. Li+/Li^0 prevents the development of high-voltage devices.

More recently, the use of phosphonium ionic liquids as an electrolyte solvent for the energy storage system has also been investigated [21–24]. MacFarlane et al. have shown the compatibility of tributylmethylphosphonium [P1444] bis(fluorosulfonyl)imide [FSI] with metallic lithium, getting a good cycling performance at 30 °C with 160 m Ah·g^{-1} for 50 cycles at 0.03 C. But these phosphonium ionic liquids may play a significant role in the curing process of epoxy networks as well as on the final properties of the resulting thermosets [25–27]. Livi et al. have shown that the chemical nature of anion has a clear impact on the polymerization of epoxy networks. Indeed, counter anions such as dicyanamide [DCA] or trimethylphosphinate [TMP] act as initiators of the anionic polymerization of the epoxy prepolymer which is due to their basic nature. For these different reasons, the design of electrolyte with good transport properties and its effect on the polymerization of epoxy-amine networks are required in this study.

The main objectives of this work were to investigate the electrochemical transport properties of the gelled electrolytes based on phosphonium ILs confined in an epoxy-amine network by electrochemical impedance spectroscopy (EIS), cyclic voltammetry (CV), and galvanostatic cycling Li | LFP cell. Thus, the transport properties of phosphonium electrolytes were performed in order to determine the ionic conductivity as well as the diffusion coefficient by Nuclear Magnetic Resonance (NMR). Then, the influence of phosphonium electrolyte on the polymerization kinetics of a rubbery epoxy network was investigated. Their thermal and thermo-mechanical behaviors were also studied by dynamical mechanical analysis (DMA) and thermo-gravimetric analyses (TGA).

2. Materials and Methods

2.1. Materials and Characterization Methods

Table 1 reviews all the structure and properties of the materials used in this study. Epoxy prepolymer based on Diglycidyl ether of polypropylene glycol (PPO) was purchased from Sigma Aldrich (St. Quentin Fallavier, France). The hardener used for the curing process is Jeffamine® D-2000 polyoxypropylene-diamine, and was supplied by Huntsman (Everslaan, Belgium). The ionic liquids (ILs) denoted trihexyl(tetradecyl)phosphonium bis(2,4,4-trimethyl-pentyl)phosphinate [P_{66614}][TMP] and tri-hexyl(tetradecyl)phosphonium bis (trifluoromethanesulfonyl) imide [P_{66614}][TFSI] were supplied by Cytec, Inc (Thorold, ON, Canada). Their structures have quaternary phosphonium cations with n-alkyl chains combined with different anions, phosphinate [TMP] and trifluoromethanesulfonimide [TFSI]. The lithium salts denoted lithium bis(2,4,4-trimethyl-pentyl)phosphinate LiTMP and lithium bis(trifluoromethanesulfonyl)imide LiTFSI were supplied by Solvionic (Toulouse, France) and Sigma Aldrich, respectively.

Table 1. Chemical structures and characteristics of the main compounds.

Name	Structure	Characteristics
PPO	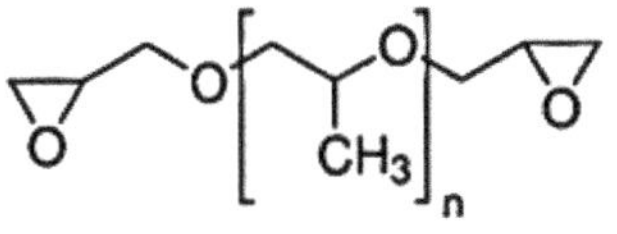	Sigma Aldrich; 320 g/eq
Jeffamine®D2000	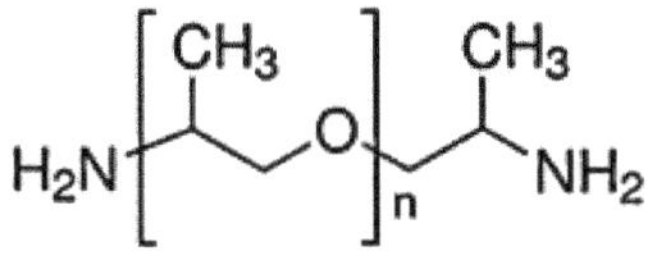	Supplied by Huntsmann; 514 g/eq
[P_{66614}][TMP]	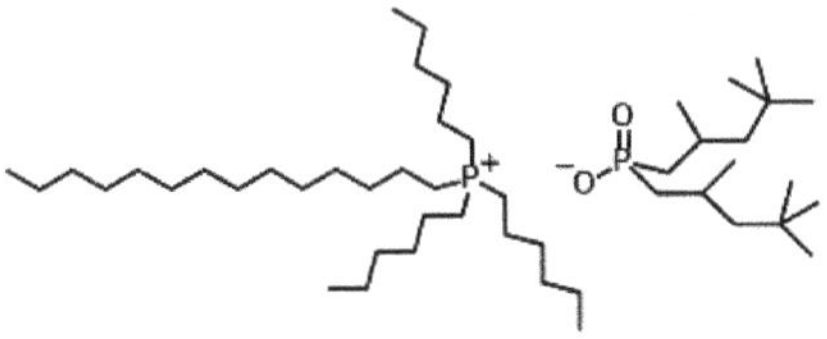	Supplied by Cytec; Molar mass = 773.27 g·mol^{-1} m.p. =−72 °C Td max = 350 °C
[P_{66614}][TFSI]	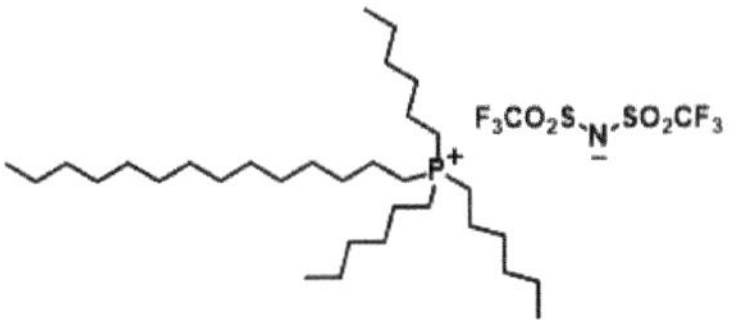	Supplied by Cytec; Molar mass = 764.0 g·mol^{-1} m.p. = −72.4 °C Td = 450 °C
LiTFSI	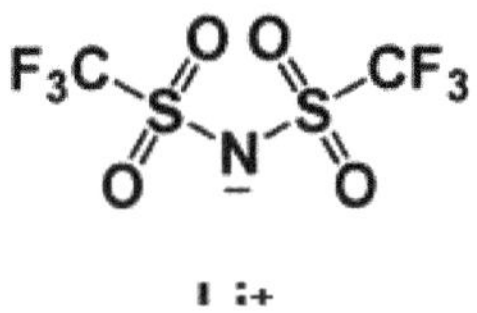	Supplied by Sigma Aldrich; Molar mass = 287.0 g·mol^{-1}
LiTMP	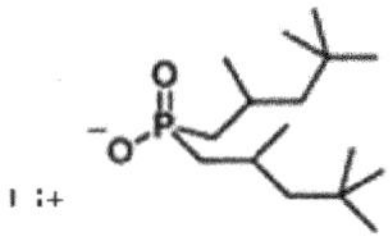	Supplied by Solvionic Molar mass = 780.27 g·mol^{-1}

Near-infrared spectroscopy (*NIR*) was recorded by using a Bruker Spectrometer (Marne la vallée, France) equipped with a heating cell to monitor changes in adsorption in the region of 10,000 to 4000 cm^{-1} with a resolution of 4 cm^{-1} on 32 scans. A temperature control was used to monitor the curing of the sample in the same conditions. The reactive mixture was injected in a glass cell with a path-length of 8 mm when the controller reached the curing temperature.

Dynamic mechanical thermal analysis was performed using a Rheometrics Solid Analyser RSA II (TA instruments, New Castle, DE, USA) at 0.01% tensile strain and a frequency of 1 Hz. The heating rate was 3 °K·min^{-1}. The temperature range was [−70 °C, 20 °C] in the case of epoxy/amine systems in stoichiometric ratio and [−70 °C, 70 °C] for the epoxy/amine systems in sub-stoichiometric ratio.

Thermogravimetric analysis (*TGA*) of Epoxy-Jeffamine/[ILs] blends were performed on a Q500 thermogravimetric analyzer (TA instruments, New Castle, DE, USA). The samples were heated from 30 to 550 °C at a rate of 20 K·min^{-1} under nitrogen flow.

Ionic conductivity was measured between aluminum electrodes by electrochemical impedance techniques using VMP 3-BioLogic (Seyssinet-Pariset, France). AC impedance measurement was measured over the frequency range from 106 Hz to 10 mHz with the potential amplitude of 50 mV. In order to investigate the temperature dependence of ionic conductivity for the electrolyte gel, the measurement was carried out on a range of temperatures between 20 and 220 °C. As the EIS spectrum did not present any semi-circle because the electrolyte was a sufficient conductor and the electrodes were blocking. Thus, the different Nyquist plots were fitted with a common model where the model circuit used is shown in the Scheme 1 as follows:

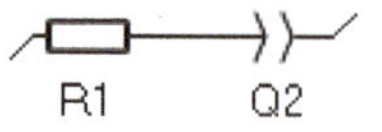

R1 Q2

Scheme 1. Model circuit used for the fitting of Nyquist plots.

where R1 corresponding to Rel.

Electrochemical stability windows (*ESWs*) of the gelled electrolyte and neat polymer were evaluated by cyclic voltammetry in a linear sweep at a sweep rate of 0.05 Mv·s^{-1} at 100 °C. The measurements were performed using the VMP 3-Biologic system.

Electrochemical stability window of gel material: Starting from Open Circuit Voltage, the ESW was measured in coin cells in oxidation by cycling at 0.05 mV/s @ 100 °C from OCV to 4 V vs. Li^+/Li for 20 cycles. A similar experiment was made in reduction, on other coin cells, by cycling at 0.05 mV/s @ 100 °C from OCV to 0.05 V vs. Li+/Li for 20 cycles.

The electrochemical performance of the Li | LFP cells was characterized using galvanostatic charge/discharge tests using VMP 3-Biologic system at the same potential range from 2.5 to 4.1 V at 100 °C.

The ^{31}P and ^{7}Li high-resolution liquid state NMR experiment were performed using an Avance DPX 400 spectrometer (Bruker, Wissembourg, France), operating at 400 MHz for ^{1}H. For the scaling of ^{31}P and ^{7}Li chemical shifts, their 0 ppm references were set up using H_3PO_4 and LiCl D2O solutions respectively.

^{1}H and ^{7}Li Pulsed Field Gradient NMR measurements were performed with an Advance Bruker spectrometer operating at 200 MHz for ^{1}H nucleus and 70.8 MHz for ^{7}Li. A multinuclear self-diffusion probe was used (Bruker Diff30 probe, gradient coil equal to 0.3 T·m^{-1}·A^{-1}). The maximum magnetic field gradient was 12 T·m^{-1}. A standard stimulated echo sequence with a diffusion time in between 10–100 ms was used [28]. The diffusion constant was obtained from the gaussian attenuation of the NMR spectrum integral versus the field gradient increase. The sample temperature controller was a Bruker BVT3000 with a+/−0.1 °C stability. Solid-state ^{7}Li NMR was carried out on the same spectrometer using a 7 mm CPMAS probe, the rotors were filled with the sample to analyze. Direct ^{7}Li excitation with high power proton decoupling during signal acquisition was applied.

2.2. Preparation of Gelled Electrolyte

The network was prepared from a mixture of amine and epoxy prepolymer with a stoichiometric ratio of 1.56. The homogenization of the mixture was carried out by magnetic stirring at 80 °C. In a first step, the epoxy-amine mixtures were precured during 6 h at 110 °C. In a second step, different contents (from 50 to 70 wt %) of phosphonium ILs were added to the precured system under mechanical stirring at 80 °C for 10 min. The solutions were degassed in an ultrasonic bath for 10 min and poured into a silicon rubber mold or coated on a PET (polyethylene terephthalate) sheet. The cure was performed at 130 °C for 6 h (to complete the polymerization).

In addition and to avoid the moisture content of the ionic liquids dramatically modifying both the ionic conductivity as well as the viscosity, all the preparation of the gelled electrolytes and the conductivity measurement have been performed in an argon-filled glove box or in coin cells assembled in an argon-filled glove box. Prior to this entry in the glove-box, the IL and the Li salt have been dried under a vacuum at 120 °C for 48 h, to remove any traces of moisture.

2.3. *Preparation of the Cathode ($LiFePO_4$) and Assembly of the Cell*

The cathode was prepared by casting a slurry of $LiFePO_4$ (Sigma-Aldrich) as an active material (45 wt %), filled with carbon black (5 wt %) and gelled electrolyte precured as a binder (50 wt %) in *N*,*N*-Dimethylformamide (DMF, Sigma-Aldrich) on an aluminum current collector. After the cure at 130 °C for 6 h and evaporation of the solvent at mild heating under vacuum (10 m Bar), the electrode disc (14 mm diameter) was an average mass coating of 2 mg·cm^{-2}.

The lithium polymer cell (Li | LFP) was assembled by placing it in a proper order: a lithium metal disc as an anode, a film of the gelled electrolyte, and the LFP as a cathode in a coin cell (CR2032). The cell system was prepared in the Argon filled glove box.

3. Results & Discussion

3.1. *Transport Properties in Phosphonium Electrolyte*

3.1.1. Ionic Conductivity

The amount of lithium salt corresponding to the maximum soluble in ionic liquid was determined to 0.75 M for LiTFSI in [P_{66614}][TFSI] and 0.2 M for LiTMP in [P_{66614}][TMP]. The ionic conductivity of phosphonium electrolytes measured by Electrochemical Impedance Spectroscopy (EIS) and calculated by Equation (1) is reported in Figure 1.

$$\sigma = \frac{d}{S.\, R_{el}} \tag{1}$$

where d is the electrolyte thickness (cm), S the activity area (cm^2), and Rel the resistance of the electrolyte (Ω).

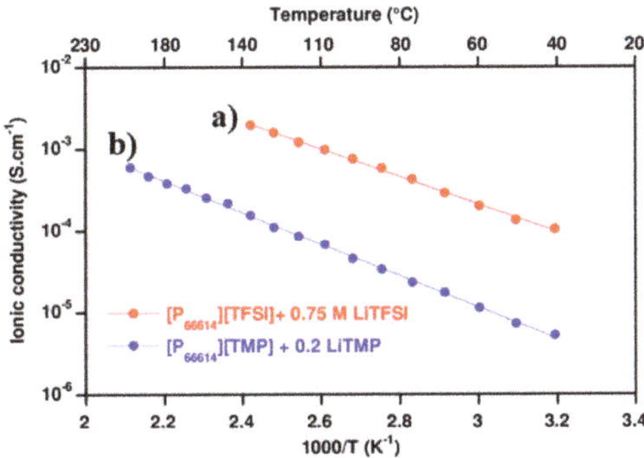

Figure 1. Ionic conductivity of phosphonium electrolytes with two types of anion: (a) [P_{66614}][TFSI] with 0.75 M of LiTFSI and (b) [P_{66614}][TMP] with 0.2 M of LiTMP.

The ionic conductivity of electrolyte [TFSI] is two decades higher compared to electrolyte [TMP] one. 1. Figure 1 shows that the log of the measured ionic conductivity is linear with respect to 1/T (the Arrhenius law). By modelling with an Arrhenius law, the activation energy could be determined and associated to the ion mobility in the electrolyte. The values obtained are summarized in Table 2. The activation energy of electrolyte [TFSI] (470 J·mol^{-1}) is lower than electrolyte [TMP] one (530

J·mol^{-1}), which can explain a better ion mobility in the electrolyte [TFSI]. In addition, this difference of ionic conductivity between [TFSI] and [TMP] based electrolytes can be explained by a higher salt concentration for electrolyte [TFSI] due to the difference in solubility. Moreover, TMP is a larger and non-fluorinated anion that has a significant influence on the mobility of the global ion in the electrolyte. The dissociation coefficient of the LiTMP is largely lower than for LiTFSI, due to the lower acidity of the corresponding acid.

Table 2. Activation energy and correlation coefficients of phosphonium electrolytes.

Sample	Ea (J·mol^{-1})	R^2
[P_{66614}][TFSI] + LiTFSI	470	0.999
[P_{66614}][TMP] + LiTMP	530	0.998

3.1.2. Diffusion Coefficient

The ionic conductivity of the electrolyte corresponds to the ion mobility present in the electrolyte. To understand the mobility of ions in the electrolyte, the diffusion coefficients of proton (D[H]) and lithium (D[Li]) were investigated by Pulsed Field Gradient NMR and are presented in Table 3.

Table 3. Diffusion coefficient (in cm^2·s^{-1}) of phosphonium electrolytes.

	T (°C)	[P_{66614}][TFSI] + LiTFSI	T (°C)	[P_{66614}][TMP] + LiTMP
D [H]	28.5	9.25×10^{-8}	54	2.3×10^{-8}
			86	1.1×10^{-7}
D [Li]	21	9.35×10^{-9}	54	No signal
			86	No signal

For the electrolyte [TFSI], the measurement of D[H] associated to the phosphonium cation is of 9.25×10^{-8} cm^2·s^{-1}. This result is in agreement with the literature [29]. MacFarlan and al. have measured two diffusion coefficients corresponding to anion and cation. The values obtained for the cation is the same order (10^{-8} cm^2·s^{-1}). The diffusion coefficient D[Li] (10^{-9} cm^2 s^{-1}) is lower than the diffusion coefficient D[H]. The lithium ion has a lower mobility in the IL [P_{66614}][TFSI]. The structuration of the electrolyte and particularly the complex steric hindrance around the lithium ions decreases its mobility compared to the mobility of the phosphonium cation. An increase of temperature would allow for getting a higher lithium mobility.

For the electrolyte [TMP], the value of D[H] is the same order as the electrolyte [TFSI] one. But, no signal is observed for the lithium ion. Two main assumptions can be made to explain this phenomenon: (i) a low dissociation of lithium salt in the electrolyte or (ii) a poor lithium salt solubility in the IL [P_{66614}][TMP].

3.1.3. Lithium Solubility

To understand the absence of lithium signal, ^{7}Li and ^{31}P NMR spectroscopy were carried out under high-resolution liquid-state NMR conditions, in a temperature range from 25 °C up to 150 °C. The recorded spectra are shown in Figure 2.

For the spectra of ^{7}Li NMR (Figure 2a), only one peak was observed, corresponding to a lithiated species. However, the chemical shifts and the peaks widths clearly change with the temperature; the linewidth becomes larger as the temperature decreases. These changes can be unambiguously assigned to the solubility feature of the lithiated species in the medium. In fact, the temperature increases the solubility of salt in the solvent. It means that between 100 °C and 25 °C the salt becomes less and less mobile due to its lower solubility. Around 25 °C, the NMR signal completely disappears under liquid-state NMR conditions. As a matter of fact, a broad line cannot be observed using the high-resolution liquid-state NMR probe. However, using solid-state NMR conditions, a ^{7}Li signal of

200 ppm (not shown in this article) could be recorded, which is fully consistent with the immobilized state of the salt. The line broadening is due to a chemical shift in the anisotropy and dipolar interactions. This observation is fully consistent with the fact that no signal was measured during the analysis of diffusion coefficient at 54 °C and 86 °C. The signal of lithium becomes finer and more intense from 130 °C.

The ^{31}P NMR spectrum determines the temperature from which the lithium salt is soluble through the phosphinate anion, which is common to lithium salt and to IL. The peak associated to the phosphonium cation (34 ppm at 25 °C) was used as a reference since it appeared unchanged regardless of the salt solubility. The second peak at 27 ppm at 25 °C corresponds at the anion phosphinate signal [30]. As for the ^{7}Li NMR, an evolution of the chemical shift of the two ^{31}P signals with the temperature can be observed. The cation peak shifts about 1.2 ppm, while a larger shift of about 3.1 ppm is observed for the anion. The temperature dependence of the chemical shift of the anion clearly presents two regimes since a significant change appears between 100 °C and 130 °C. That could be assigned to a chemical environment change of the anion associated to the better solubility of the lithium salt above 100 °C.

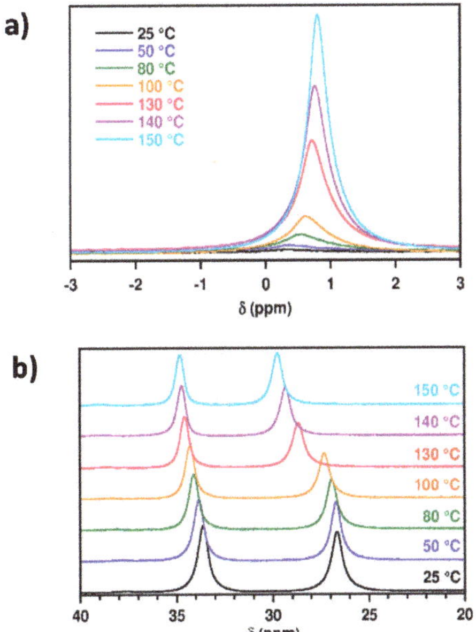

Figure 2. ^{7}Li NMR (**a**) and ^{31}P NMR (**b**) of electrolyte [TMP] ([P_{66614}][TMP] + 0.2 M LiTMP) at different temperatures.

The relative intensity of the anion peak compared to of the relative intensity of the cation peak (considered as reference) as a function of temperature is presented in Figure 3. The relative intensity of the anion peak is higher than 1 from 100 °C, which indicates the beginning of the lithium salt solubilization in the medium. For an optimal solubilization, it was necessary to use an operating temperature around 150 °C. This is a relatively high temperature for using an electrolyte in a lithium battery.

To conclude this part, the electrolyte [TFSI] presents a better ability to be used as an electrolyte in a lithium battery. Therefore, this gelled electrolyte [TFSI] was chosen and its properties were investigated. Specifically, the electrochemical performance in the Li | LFP cell was discussed.

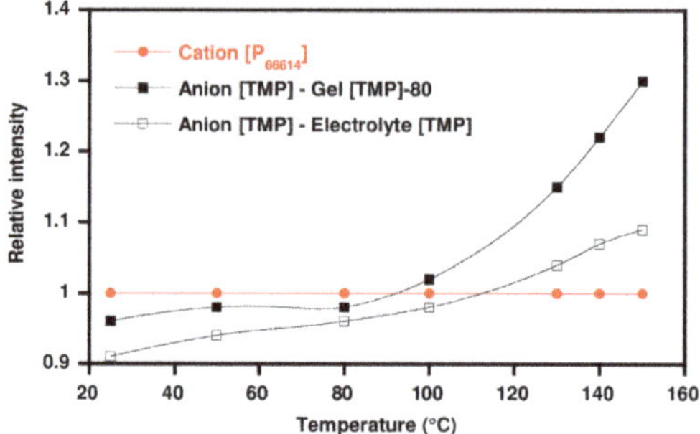

Figure 3. Evolution of the normalized relative intensity of the phosphinate anion compared to the relative intensity of the cation as a function of the temperature for the electrolyte [TMP] ([P_{66614}][TMP] + 0.2 M LiTMP).

3.2. *IL Confinement within the Epoxy Network*

3.2.1. Effect of IL Content on the Exudation

Table 4 presents the exudation behavior of cured epoxy networks modified with varying electrolyte content. The electrolyte exhibits a high miscibility since a maximum of ionic liquid of 70 wt % can be incorporated in a gel state. However, when more electrolyte is added, the sample becomes difficult to handle.

Table 4. Name and composition of different gelled electrolytes.

Sample	Composition (wt %)			Exudation
	Polymer	IL	LiTFSI	
PPO [TFSI]-50	50	41.0	9.0	No
PPO [TFSI]-60	40	49.2	10.8	No
PPO [TFSI]-65	35	53.3	11.7	No
PPO [TFSI]-70	30	57.4	12.6	No

3.2.2. Effect of IL on Reaction Progress

A NIR analysis was investigated to understand the differences in the reaction mechanism within an epoxy prepolymer-amine system with and without ionic liquids. In all the cases, the curing procedure was kept the same; i.e., 6 h at 110 °C and 6 h at 120 °C. The different band assignments are listed in Table 5 [31]. At the beginning, the spectrum of the initial mixture displays an absorption peak of the epoxy group at 4530 cm^{-1}, an absorption peak of the primary amine group at 4935 cm^{-1}, and an absorption peak of secondary and primary amine groups at 6500 cm^{-1}. The evolution of these peaks follows the reaction, particularly after the gel time.

The cross-linking process is quantified from the decrease of these absorption peaks at 4530 and 4935 cm^{-1}. The conversion (α) of epoxy group is calculated from the relationship (2) [31].

$$\alpha_{epoxy,t} = 1 - \frac{(A_{epoxy,4530})_t}{(A_{epoxy,4530})_{t=0}} \quad (2)$$

where A_{epoxy} are the areas of the epoxy peak at the curing time (t) and at the start of the reaction (t = 0).

Equation (3) can also be applied to calculate the conversion of the primary amine characterized by the band at 4935 cm^{-1}:

$$\alpha_{PA,t} = 1 - \frac{(A_{PA,4935})_t}{(A_{PA,4935})_{t=0}} \quad (3)$$

where PA represents the primary amine (PA) absorption band at 4935 cm^{-1}.

Equation (4) permits to follow the evolution of the both primary and secondary amine functions (A) at 6500 cm^{-1}:

$$\alpha_{A,t} = 1 - \frac{(A_{A,6500})_t}{(A_{A,6500})_{t=0}} \tag{4}$$

where A represents the primary and secondary amine (A) absorption band at 6500 cm^{-1}. The evolution of secondary amine can be determined with Equation (5):

$$\beta_{SA,t} = \alpha_{PA,t} - \alpha_{A,t} \tag{5}$$

Table 5. Near infrared (IR) Band Assignments of PPO-Jeffamine polymerization.

Wavenumber (cm^{-1})	Peak Assignment
7099	O–H overtone
6600–6480	Primary and Secondary amine combination band Overtones of N–H stretching
6072	Terminal epoxy first overtone of C–H stretching
5880–5500	C–H overtone (CH_2, CH_3)
5249	–OH due to moisture (O–H asymmetric stretching and bending)
4935	Primary amine combination band N–H stretching and bending
4530	Epoxy combination band (C–H stretching and epoxy ring breathing)

Thus the conversion of the epoxide groups (E), primary amine (PA), and the evolution of secondary amine versus reaction time were plotted in Figure 4 for 50 wt % of ([P_{66614}][TFSI + 0.75 M LiTFSI). The kinetic modelling of epoxy and primary amine functions conversion is made according to a first-order kinetic law according to Equation (6):

$$\frac{dx}{dt} = k(1-x)x = x_{\infty} - x_0 \exp^{-kt} \tag{6}$$

where x is the conversion, k is the kinetic constant of the reaction, x_{∞} is the conversion at an infinite time (usually equal to 1), x_0 is the conversion at t = 0 or t = 6 h. The kinetic constants from the modelling are shown in Table 6.

For the neat PPO-Jeffamine network, during the first step of curing, the reaction between the epoxy and primary amine groups is favored since the kinetic constant associated to the primary amine is higher than the one corresponding to the secondary amine (kAP = 0.534 $h^{-1/2}$). The temperature increase results in the kinetic constants of the epoxy and secondary amines function higher. The function conversion does not reach 100% along the curing process because the temperature is limited at 120 °C during the second curing cycle.

The addition of the 50 wt % electrolyte induces a strong increase of kinetic constants of epoxy, as well as the secondary and primary amine functions. A total conversion of the amine and epoxy functions is reached at the end of the curing process. Therefore, the addition of electrolyte has a catalytic effect on the polymerization of the PPO-Jeffamine network. These results are in agreement with literature [32,33] that proves lithium salt LiTFSI, which is a lewis acid, helps the opening of the epoxy ring and thus catalyzes the polymerization of the epoxy amine network. This effect is different compared to our previous works where the anions of ionic liquids initiated the epoxy etherification [27].

In summary, it was shown that the ionic liquid [P_{66614}][TFSI] presents a good compatibility with the network. The NIR analysis for monitoring the polymerization kinetics highlights the effect of the anion as a catalyzer of PPO-Jeffamine polymerization. Let us now consider the influence of the electrolyte on the final properties of the epoxy networks.

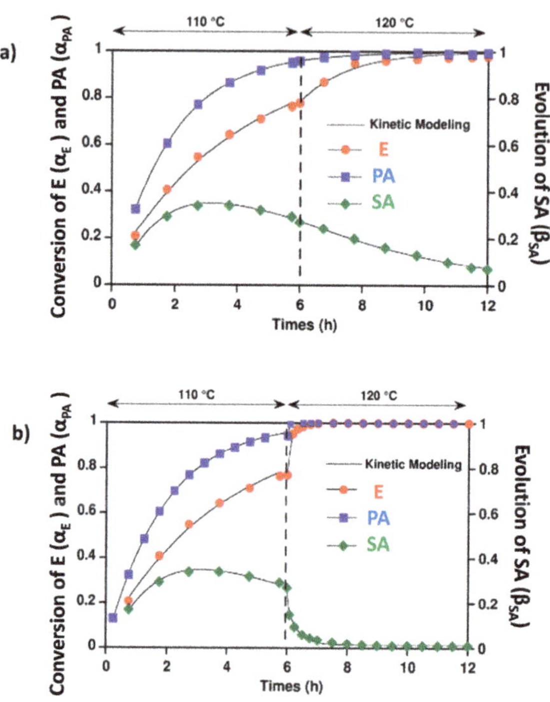

Figure 4. Epoxy conversion (E), and primary amine (PA) and evolution of secondary amine (SA) during the reaction time at 110 °C for 6 h and at 120 °C for 6 h of (**a**) neat PPO-Jeffamine networks compared to (**b**) PPO-Jeffamine/Electrolyte system with 50 wt % of ([P_{66614}][TFSI] + 0.75 M LiTFSI).

Table 6. The kinetic constants ($h^{-1/2}$) of reaction for epoxy functions (k_E), for primary amine functions (k_{PA}) and for secondary amine functions (k_{SA}) and correlation coefficients.

System	PPO-Jeffamine		PPO-Jeffamine + ([P_{66614}][TFSI] + LiTFSI)	
Step of Curing	**First Step**	**Second Step**	**First Step**	**Second Step**
k_E ($h^{-1/2}$)	0.272	0.988	0.268	7.989
(R^2)	(0.994)	(0.996)	(0.989)	(0.993)
k_{PA} ($h^{-1/2}$)	0.534	0.534	0.532	30.266
(R^2)	(0.999)	(0.999)	(0.999)	(0.999)
k_{SA} ($h^{-1/2}$)	0.190	0.243	0.186	3.039
(R^2)	(0.988)	(0.996)	(0.990)	(0.990)

3.3. *Influence of the Electrolyte on the Final Properties of Epoxy Based-Networks*

3.3.1. On the Thermal Stability of Networks

The thermal stability of PPO-Jeffamine/Electrolyte networks was investigated by thermogravimetric analysis (TGA) and compared to a neat epoxy-amine network and pure electrolyte. Figure 5 gives the weight loss and the corresponding derivative (DTG) of the networks as a function of the electrolyte content.

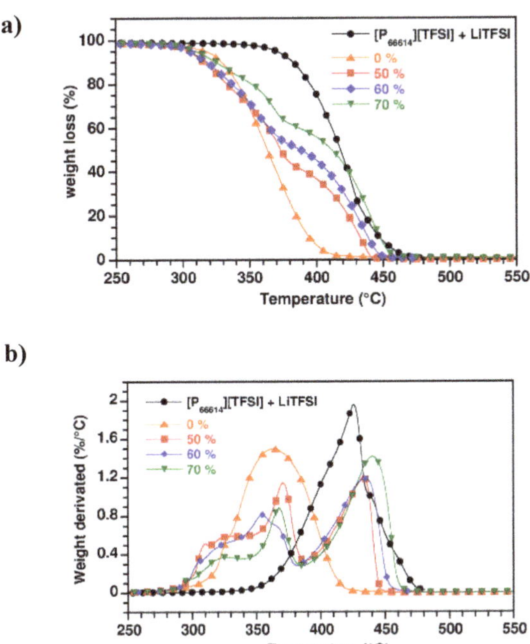

Figure 5. Thermo-gravimetric analyses (TGA) (**a**) and dynamical mechanical analysis (DTG) (**b**) of electrolyte ($[P_{66614}][TFSI]$ + 0.75 M LiTFSI) alone and introduced in PPO-Jeffamine networks prepared with an electrolyte varying content (heating ramp: 10 $K \cdot min^{-1}$; atmosphere: Nitrogen).

In Figure 5, the thermal stability of the electrolyte (Td5% = 370 °C) is much higher than the epoxy network (Td5% = 319 °C). The gelled electrolytes have an intermediate thermal stability between one of the host networks and the electrolyte one, depending on the electrolyte content. The DTG (Figure 5b) clearly identifies two degradation peaks corresponding to the degradation of the network between 300 and 385 °C and the second one to the degradation of the electrolyte localized between 385 and 460 °C. All the gelled samples present a high thermal stability but are limited by the epoxy network (Td5% = 310 °C).

3.3.2. On the Viscoelastic Properties

The effect of electrolyte content on the viscoelastic properties of PPO-Jeffamine networks was investigated by dynamical mechanical analysis. The relaxation temperature ($T\alpha$), defined as the maximum value of tan δ and the storage modulus E' at rubbery state at -10 °C are summarized in Table 7.

In both cases, one single peak of relaxation was observed, suggesting a homogeneous network without a macroscale phase separation phenomenon. Then, the addition of the electrolyte induces systematically a slight increase of $T\alpha$ from -49 °C to -43.0 °C in the presence of 70 wt % of electrolyte. On the other side, the addition of the electrolyte leads to a strong decrease of storage modulus E' at a rubbery state, which can be attributed at a decrease of network crosslink density [34,35]. Thus, E' decreases 80% in the presence of 70 wt % of electrolyte. So, the addition of a large amount of electrolyte in the gel significantly impacts the mechanical strength of the gelled electrolyte by increasing the gel flexibility and making it even more difficult to handle. This observation can be correlated at the diluted effect of the medium, where the probability of the prepolymers to meet each other decreases with the addition of the electrolyte. The semi-diluted medium then promotes intra-molecular reactions like cyclization are considered defective and increase the apparent elasticity of the network.

Table 7. DMA data of PPO-Jeffamine/Electrolyte networks.

Electrolyte Content (wt %)	Tα (°C)	Rubbery State E′ (MPa)
0	−49.2	0.33
50	−48.8	0.25
60	−46.9	0.20
70	−43.0	0.06

3.4. *Effect of Electrolyte on the Transport and Electrochemical Properties*

3.4.1. Transport Properties

To confirm the potential use of this system as a gelled electrolyte for lithium-ion battery, a film with low thickness (<100 μm) was prepared by solvent casting. The gelled electrolytes were prepared with different amounts of electrolyte containing $[P_{66614}]$[TFSI]/lithium salt (LiTFSI). The ionic conductivities were measured by Electrochemical Impedance Spectroscopy (EIS) and calculated by Equation (1).

Figure 6 shows the temperature dependence (in the range of 20–140 °C) of the ionic conductivities of the gelled electrolyte with different amounts of electrolyte in comparison with the neat electrolyte. The ionic conductivity of the gelled electrolyte increases both with the amount of electrolyte and with the temperature with a maximum value of 0.13 $mS \cdot cm^{-1}$ at 100 °C for 70 wt % electrolyte based gel. This phenomenon results from a higher number of ions and faster ion movements since it is thermally activated and leads to higher ionic conductivity. The ionic conductivity of the gelled electrolyte is lower than the neat electrolyte, because of the decrease of ion mobility confined in the host network.

By modelling the ionic conductivity plots with an Arrhenius law, the activation energy values were determined and reported in Table 8. These Ea values are independent on the electrolyte content because only the electrolyte trapped between the polymer chains is responsible for the ionic conduction.

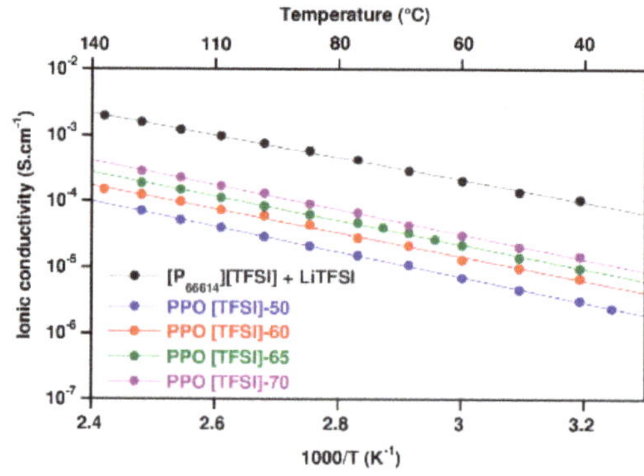

Figure 6. Ionic conductivity of gelled electrolyte with different amount of electrolyte ($[P_{66614}]$[TFSI] with 0.75 M of LiTFSI).

Table 8. Activation energy and correlation coefficients of gelled electrolyte.

Sample	Ea ($J \cdot mol^{-1}$)	R^2
[P_{66614}][TFSI] + LiTFSI	440	0.999
PPO [TFSI]-50	500	0.999
PPO [TFSI]-60	470	0.997
PPO [TFSI]-65	475	0.999
PPO [TFSI]-70	490	0.999

3.4.2. Electrochemical Stability

To confirm the electrochemical stability of the gelled electrolyte, a cyclic voltammetry (CV) has been used to know this electrochemical stability window. The CV curves are shown in Figure 7. The measurement has been performed in coin cells. The working electrode is a stainless steel disk. The counter electrode is metallic lithium, acting as well as a reference electrode.

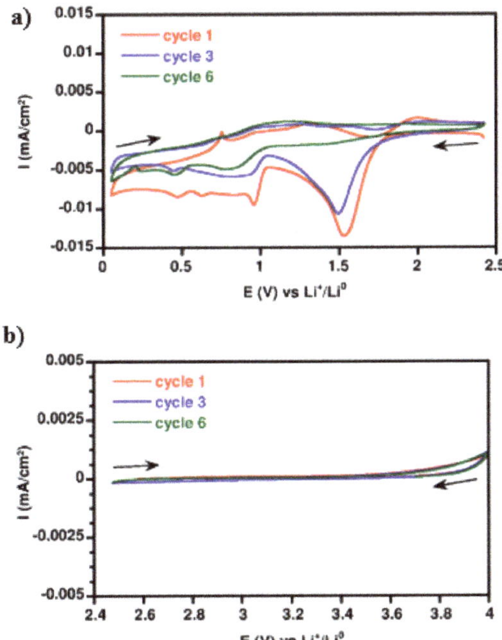

Figure 7. Cyclic voltammetry of gelled electrolyte prepared with 70 wt % of electrolyte (**a**) in reduction to 0.05 V vs. Li^+/Li^0 and (**b**) in oxidation to 4 V vs. Li^+/Li^0.

In the reduction (Figure 7a), the electrochemical phenomenon at 1.5 V and 1 V are observed with a current intensity measured at -0.014 $mA \cdot cm^{-2}$ and at 0.01 $mA \cdot cm^{-2}$, respectively, for the first cycle. This phenomenon, observed as well for the electrolyte [TFSI], was attributed to the formation of a passivation layer that stabilizes at the 6th cycle. As the electrolyte is based on ionic liquids, the SEI should be different, as in conventional carbonate based electrolytes. The nature of the passivation layer is attributed to the chemical nature of the IL. In oxidation (Figure 7b), the current measured is very low (<2 $\mu A \cdot cm^{-2}$). Thus, it is possible to consider the gelled electrolyte is electrochemically stable with stainless steel to 4 V vs. $Li+/Li^0$.

The cathodic and anodic limiting potentials are about 0.05 V and 4 V vs. $Li+/Li^0$, respectively, and ESW is 3.95. These results suggest that gelled electrolyte PPO [TFSI]−70 is a good candidate to be used as an electrolyte in lithium ion batteries with the metallic lithium anode and the $LiFePO_4$, with an active plateau at about 3.4 V vs. Li^+/Li^{0}[3].

3.4.3. Cycling Performance of Li | $LiFePO_4$ Cell

The gelled electrolyte was tested in the Li | $LiFePO_4$ cell at different current densities (C/100, C/50, and C/20) between 2.5 V and 4 V at 100 °C. The results are presented for the cycle 3, 7 and 15 in Figure 8.

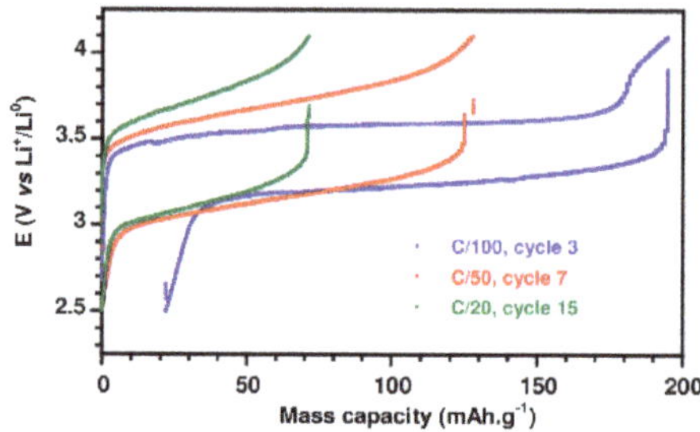

Figure 8. Cycling performances of Li | $LiFePO_4$ cell at different current densities (C/100, C/50 and C/20) between 2.5 V and 4 V at 100 °C of gelled electrolyte prepared with 70 wt % of electrolyte.

The charge/discharge curves for the cell under a low current rate (C/100) are relatively flat lines at about 3.3 and 3.5 V, respectively. The curves obtained for the higher current rates (C/50 and C/20) deviate from this linear behavior. This can be explained by a lower ion diffusion within the electrolyte and at the electrode/electrolyte interface.

The Li | PPO [TFSI]−70 | $LiFePO_4$ shows a good cycling performance, with large charge/discharge capacities ranging from 210 m $Ah{\cdot}g^{-1}$ (slightly upper of the theoretical capacity) at C/100 and the initial coulombic efficiency is ~90% (Figure 9). The additional current consumption in charge was attributed to the degradation of the gelled electrolyte. When the current rate was increased to C/50, and then to C/20, the charge and discharge capacity was approximately the same with a better coulombic efficiency. However, the specific capacity was decreased to around 120 m $Ah{\cdot}g^{-1}$ then to 70 m $Ah{\cdot}g^{-1}$, which corresponds to 40% of the theoretical capacity. This decrease could be correlated to the lower ionic conductivity of the gelled electrolyte, which could induce an additional ohmic resistance and concentration gradients within the electrolyte. Therefore, the charge transfer could be more difficult to the electrodes and the polarization of the cell would be increased. Nevertheless, a good cycle stability was observed at C/20 over 20 cycles with a capacity of around 70 m $Ah{\cdot}g^{-1}$.

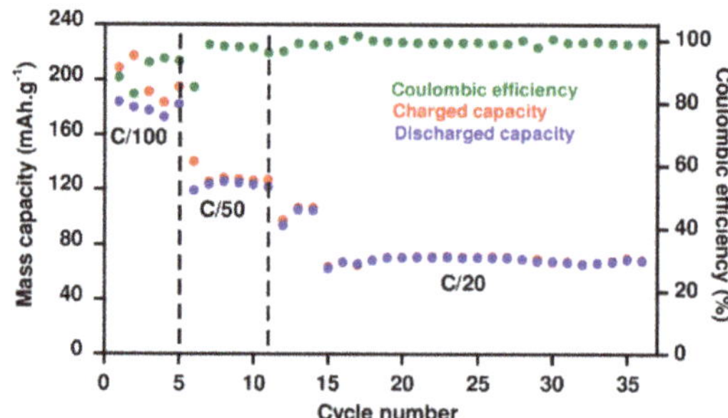

Figure 9. Cycling performance of the Li | PPO [TFSI]−70 | $LiFePO_4$ cell at indicating rates at 100 °C.

4. Conclusions

In this work, the transport properties of different phosphonium electrolytes denoted [P_{66614}][TMP] + LiTMP and [P_{66614}][TFSI] LiTFSI were investigated. We have shown that the electrolyte [TMP] have a low solubility of the lithium salt at ambient temperature. The NMR analysis determines the operating temperature at 150 °C. This temperature is too high to consider its use as an electrolyte for lithium battery. However, the electrolyte [TFSI] has shown a better ability for this use.

So, the electrolyte [TFSI] was confined within epoxy-amine networks. The influence of the electrolyte on the polymerization kinetics as well as on the final properties (thermal and mechanical) of the epoxy networks was performed. Thus, we have demonstrated that the electrolyte [TFSI] catalyzed the polymerization of the epoxy/amine prepolymer. We have highlighted the ability to develop flexible epoxy networks swollen with high amounts of electrolyte [TFSI] (70 wt %).

The gelled electrolyte has an ionic conductivity of 0.13 $mS{\cdot}cm^{-1}$ at 100 °C with an electrochemical stability window at 3.95 V. The test of cycling performance has shown for the first time the possibility to use a phosphonium gelled electrolyte in the Li | LFP cell. A low current rate was necessary to obtain a theoretical capacity of the material. Thus, a proof of concept has been achieved. In summary, this system could find some applications in extreme conditions, as very high temperature use requires high safety protocols. For conventional applications, a modification of the system (shorter IL, different polymer) is required to fit the expected performances as well as an optimization of the cell and the understanding that some phenomena involved at the interface of the electrolyte/electrode require further study.

Author Contributions: J.D.-R., S.L., L.P. conceived the paper and designed the experiments. M.L., L.B., A.G., M.B. performed the experiments. M.L., J.D.-R., S.L. analyzed the data and also wrote the paper.

Funding: This research received no external funding.

Conflicts of Interest: The authors declare no conflict of interest.

References

1. Larcher, D.; Tarascon, J.M. Towards greener and more sustainable batteries for electrical energy storage. *Nat. Chem.* **2015**, *7*, 19–29. [CrossRef] [PubMed]
2. Bruce, P.G.; Freunberger, S.A.; Hardwick, L.J.; Tarascon, J.M. Li–O_2 and Li–S batteries with high energy storage. *Nat. Mater.* **2012**, *11*, 19–29. [CrossRef] [PubMed]
3. Padhi, A.K.; Nanjundaswamy, K.S.; Goodenough, J.B. Phospho-olivines as positive-electrode materials for rechargeable lithium batteries. *J. Electrochem. Soc.* **1997**, *144*, 1188–1194. [CrossRef]
4. Ohzuku, T.; Ueda, A.; Yamamoto, N. Zero-Strain Insertion Material of Li [$Li_{1/3}Ti_{5/3}$] O_4 for Rechargeable Lithium Cells. *J. Electrochem. Soc.* **1995**, *142*, 1431–1435. [CrossRef]
5. Armand, M.; Endres, F.; MacFarlane, D.R.; Ohno, H.; Scrosati, B. Ionic-liquid materials for the electrochemical challenges of the future. *Nat. Mater.* **2009**, *8*, 621–629. [CrossRef] [PubMed]
6. Passerini, S.; Montanino, M.; Appetecchi, G.B. Lithium Polymer Batteries Based on Ionic Liquids. In *Polymer for Energy Storage and Conversion*; John Wiley & Sons, Inc.: Hoboken, NJ, USA, 2013; p. 53.
7. Ferrari, S.; Quartarone, E.; Mustarelli, P.; Magistris, A.; Fagnoni, M.; Protti, S.; Gerbaldi, C.; Spinella, A. Lithium ion conducting PVdF-HFP composite gel electrolytes based on *N*-methoxyethyl-*N*-methylpyrrolidinium bis (trifluoromethanesulfonyl)-imide ionic liquid. *J. Power Source* **2010**, *195*, 559–566. [CrossRef]
8. Earle, M.J.; Seddon, K.R. Ionic liquids. Green solvents for the future. *Pure Appl. Chem.* **2000**, *72*, 1391–1398. [CrossRef]
9. Tokuda, H.; Ishii, K.; Susan, M.A.; Tsuzuki, S.; Hayamizu, K.; Watanabe, M. Physicochemical properties and structures of room-temperature ionic liquids. 3. Variation of cationic structures. *J. Phys. Chem. B* **2006**, *110*, 2833–2839. [CrossRef] [PubMed]
10. Olivier-Bourbigou, H.; Magna, L. Ionic liquids: Perspectives for organic and catalytic reactions. *J. Mol. Catal. A Chem.* **2002**, *182*, 419–437. [CrossRef]

11. Lane, G.H. Electrochemical reduction mechanisms and stabilities of some cation types used in ionic liquids and other organic salts. *Electrochim. Acta* **2012**, *83*, 513–528. [CrossRef]
12. Kim, G.T.; Appetecchi, G.B.; Alessandrini, F.; Passerini, S. Solvent-free, PYR1ATFSI ionic liquid-based ternary polymer electrolyte systems: I. Electrochemical characterization. *J. Power Source* **2007**, *171*, 861–869. [CrossRef]
13. Kim, G.T.; Jeong, S.S.; Xue, M.Z.; Balducci, A.; Winter, M.; Passerini, S.; Alessandrini, F.; Appetecchi, G.B. Development of ionic liquid-based lithium battery prototypes. *J. Power Source* **2012**, *199*, 239–246. [CrossRef]
14. Yang, P.; Cui, W.; Li, L.; Liu, L.; An, M. Characterization and properties of ternary P (VdF-HFP)-LiTFSI-EMITFSI ionic liquid polymer electrolytes. *Solid State Sci.* **2012**, *14*, 598–606. [CrossRef]
15. Appetecchi, G.B.; Kim, G.T.; Montanino, M.; Carewska, M.; Marcilla, R.; Mecerreyes, D.; De Meatza, I. Ternary polymer electrolytes containing pyrrolidinium-based polymeric ionic liquids for lithium batteries. *J. Power Source* **2010**, *195*, 3668–3675. [CrossRef]
16. Gerbaldi, C.; Nair, J.R.; Ahmad, S.; Meligrana, G.; Bongiovanni, R.; Bodoardo, S.; Penazzi, N. UV-cured polymer electrolytes encompassing hydrophobic room temperature ionic liquid for lithium batteries. *J. Power Source* **2010**, *195*, 1706–1713. [CrossRef]
17. Stepniak, I. Compatibility of poly (bisAEA4)-LiTFSI–MPPipTFSI ionic liquid gel polymer electrolyte with $Li_4Ti_5O_{12}$ lithium ion battery anode. *J. Power Source* **2014**, *247*, 112–116. [CrossRef]
18. Stepniak, I.; Andrzejewska, E.; Dembna, A.; Galinski, M. Characterization and application of *N*-methyl-*N*-propylpiperidinium bis (trifluoromethanesulfonyl) imide ionic liquid–based gel polymer electrolyte prepared in situ by photopolymerization method in lithium ion batteries. *Electrochim. Acta* **2014**, *121*, 27–33. [CrossRef]
19. Shirshova, N.; Bismarck, A.; Carreyette, S.; Fontana, Q.P.; Greenhalgh, E.S.; Jacobsson, P.; Johansson, P.; Marczewski, M.J.; Kalinka, G.; Kucernak, A.R.; et al. Structural supercapacitor electrolytes based on bicontinuous ionic liquid–epoxy resin systems. *J. Mater. Chem. A* **2013**, *1*, 15300–15309. [CrossRef]
20. Sotta, D.; Bernard, J.; Sauvant-Moynot, V. Application of electrochemical impedance spectroscopy to the study of ionic transport in polymer-based electrolytes. *Prog. Organ. Coat.* **2010**, *69*, 207–214. [CrossRef]
21. Frackowiak, E.; Lota, G.; Pernak, J. Room-temperature phosphonium ionic liquids for supercapacitor application. *Appl. Phys. Lett.* **2005**, *86*, 164104. [CrossRef]
22. Tsunashima, K.; Yonekawa, F.; Sugiya, M. Lithium secondary batteries using a lithium nickelate-based cathode and phosphonium ionic liquid electrolytes. *Electrochem. Solid-State Lett.* **2009**, *12*, A54–A57. [CrossRef]
23. Armel, V.; Velayutham, D.; Sun, J.; Howlett, P.C.; Forsyth, M.; MacFarlane, D.R.; Pringle, J.M. Ionic liquids and organic ionic plastic crystals utilizing small phosphonium cations. *J. Mater. Chem.* **2011**, *21*, 7640–7650. [CrossRef]
24. Jin, L.; Howlett, P.C.; Pringle, J.M.; Janikowski, J.; Armand, M.; MacFarlane, D.R.; Forsyth, M. An organic ionic plastic crystal electrolyte for rate capability and stability of ambient temperature lithium batteries. *Energy Environ. Sci.* **2014**, *7*, 3352–3361. [CrossRef]
25. Silva, A.A.; Livi, S.; Netto, D.B.; Soares, B.G.; Duchet, J.; Gérard, J.F. New epoxy systems based on ionic liquid. *Polymer* **2013**, *54*, 2123–2129. [CrossRef]
26. Livi, S.; Silva, A.A.; Thimont, Y.; Nguyen, T.K.; Soares, B.G.; Gérard, J.F.; Duchet-Rumeau, J. Nanostructured thermosets from ionic liquid building block–epoxy prepolymer mixtures. *RSC Adv.* **2014**, *4*, 28099–28106. [CrossRef]
27. Leclère, M.; Livi, S.; Maréchal, M.; Picard, L.; Duchet-Rumeau, J. The properties of new epoxy networks swollen with ionic liquids. *RSC Adv.* **2016**, *6*, 56193–56204. [CrossRef]
28. Tanner, J.E. Use of the stimulated echo in NMR diffusion studies. *J. Chem. Phys.* **1970**, *52*, 2523–2526. [CrossRef]
29. Pozo-Gonzalo, C.; Howlett, P.C.; Hodgson, J.L.; Madsen, L.A.; MacFarlane, D.R.; Forsyth, M. Insights into the reversible oxygen reduction reaction in a series of phosphonium-based ionic liquids. *Phys. Chem. Chem. Phys.* **2014**, *16*, 25062–25070. [CrossRef] [PubMed]
30. Dharaskar, S.A.; Wasewar, K.L.; Varma, M.N.; Shende, D.Z.; Tadi, K.K.; Yoo, C.K. Synthesis, characterization, and application of novel trihexyl tetradecyl phosphonium bis (2, 4, 4-trimethylpentyl) phosphinate for extractive desulfurization of liquid fuel. *Fuel Process. Technol.* **2014**, *123*, 1–10. [CrossRef]

31. Pramanik, M.; Mendon, S.K.; Rawlins, J.W. Disecondary amine synthesis and its reaction kinetics with epoxy prepolymers. *J. Appl. Polym. Sci.* **2012**, *126*, 1929–1940. [CrossRef]
32. Sotta, D. Liquides Ioniques Gélifiées Pour les Batteries Lithium-Ion. Ph.D. Thesis, Université de Picardie Jules Verne, Amiens, France, 2011.
33. Cossy, J.; Bellosta, V.; Hamoir, C.; Desmurs, J.R. Regioselective ring opening of epoxides by nucleophiles mediated by lithium bistrifluoromethanesulfonimide. *Tetrahedron Lett.* **2002**, *43*, 7083–7086. [CrossRef]
34. Dušek, K.; Dušková-Smrčková, M. Polymer Networks. In *Macromolecular Engineering*; Wiley-VCH Verlag GmbH & Co. KGaA: Weinheim, Germany, 2007; p. 1687.
35. Gennes, P.G. *Scaling Concepts in Polymer Physics*; Cornell University Press: Ithaca, NY, USA, 1979.

Article

Ionic Liquids as Delaminating Agents of Layered Double Hydroxide during In-Situ Synthesis of Poly (Butylene Adipate-*co*-Terephthalate) Nanocomposites

Hynek Beneš [1,*], Jana Kredatusová [1], Jakub Peter [1], Sébastien Livi [2], Sonia Bujok [1], Ewa Pavlova [1], Jiří Hodan [1], Sabina Abbrent [1], Magdalena Konefał [1] and Petra Ecorchard [3]

1 Institute of Macromolecular Chemistry of the Czech Academy of Sciences, Heyrovského nám. 2, 162 06 Prague 6, Czech Republic; jana.kredatusova@email.cz (J.K.); peter@imc.cas.cz (J.P.); bujok@imc.cas.cz (S.B.); pavlova@imc.cas.cz (E.P.); hodan@imc.cas.cz (J.H.); abbrent@imc.cas.cz (S.A.); magdalenakonefal@imc.cas.cz (M.K.)

2 Université de Lyon, CNRS, UMR 5223, Ingénierie des Matériaux Polymères, INSA Lyon, F-69621 Villeurbanne, France; sebastien.livi@insa-lyon.fr

3 Institute of Inorganic Chemistry of the Czech Academy of Sciences, Husinec-Řež č.p. 1001, 25068 Řež, Czech Republic; ecorchard@iic.cas.cz

* Correspondence: benesh@imc.cas.cz; Tel.: +420-296-809-313

Received: 14 March 2019; Accepted: 10 April 2019; Published: 16 April 2019

Abstract: Currently, highly demanded biodegradable or bio-sourced plastics exhibit inherent drawbacks due to their limited processability and end-use properties (barrier, mechanical, etc.). To overcome all of these shortcomings, the incorporation of lamellar inorganic particles, such as layered double hydroxides (LDH) seems to be appropriate. However, LDH delamination and homogenous dispersion in a polymer matrix without use of harmful solvents, remains a challenging issue, which explains why LDH-based polymer nanocomposites have not been scaled-up yet. In this work, LDH with intercalated ionic liquid (IL) anions were synthesized by a direct co-precipitation method in the presence of phosphonium IL and subsequently used as functional nanofillers for in-situ preparation of poly (butylene adipate-*co*-terephthalate) (PBAT) nanocomposites. The intercalated IL-anions promoted LDH swelling in monomers and LDH delamination during the course of in-situ polycondensation, which led to the production of PBAT/LDH nanocomposites with intercalated and exfoliated morphology containing well-dispersed LDH nanoplatelets. The prepared nanocomposite films showed improved water vapor permeability and mechanical properties and slightly increased crystallization degree and therefore can be considered excellent candidates for food packaging applications.

Keywords: poly (butylene adipate-*co*-terephthalate); ionic liquid; layered double hydroxide; in-situ polymerization; nanocomposite; permeability; biodegradable polymer

1. Introduction

Currently, engineering plastics are designed as sustainable materials with an adjustable lifetime in order to minimize their impact on the environment. The challenging issue is thus to prepare advanced plastics either with (i) increased durability or (ii) easy and fast degradability by means of abiotic factors (humidity, temperature, pH, etc.) and/or biological (microbial) attack. The first are hi-performance materials with a long service life, while the latter are often designed as biodegradable polymers for packaging with a typical life-time shorter than one year.

Poly (butylene adipate-*co*-terephthalate) (PBAT) is a typical representative of aliphatic-aromatic copolyesters, which provide balance between biodegradability of aliphatic segments and thermo-mechanical properties of aromatic units. PBAT is a flexible material with high elongation at break and thus suitable for coatings and film applications [1]. However, its low crystallization degree results in insufficient stiffness and modulus. Together with low water vapor barrier properties, these are the main drawbacks limiting the PBAT usage for food packaging [2,3]. A possibility to overcome the PBAT shortcomings is to incorporate lamellar inorganic particles, such as cationic clays (talc and montmorillonites [2,4–7]), anionic clays such as layered double hydroxides (LDH) [8], and carbon compounds such as graphite or graphene-based nanoplatelets [9,10] in the PBAT matrix.

LDH are particularly advantageous due to their mild and efficient synthesis, chemical structure versatility and anionic exchange ability, which enables tuning of the LDH surface interactions for targeted nanocomposite fabrication [11]. The LDH or hydrotalcite-like compounds are represented by the general formula $[M^{2+}{}_{1-x}M^{3+}{}_{x}(OH)_2]^{x+}(A^{n-})_{x/n}\cdot mH_2O$, where M^{2+} and M^{3+} are divalent (Ca^{2+}, Mg^{2+}, Fe^{2+}, Zn^{2+}, Cu^{2+}, Co^{2+}) and trivalent (Fe^{3+}, Al^{3+}, Bi^{3+}, In^{3+}, La^{3+}, Ga^{3+}) layer cations, respectively, and A^{n-} is the exchangeable inorganic anion (Cl^-, OH^-, $CO_3{}^{2-}$, $SO_4{}^{2-}$, $NO_3{}^-$) in the interlayer space with x being the relative substitution rate generally ranging between $0.20 < x < 0.33$ resulting in M^{2+}/M^{3+} ratios of 2/1–4/1 [12,13].

The layered inorganic compounds can act as nano-sized fillers in a polymer matrix providing high reinforcing and barrier effects only when their structure is delaminated [14]. Additionally, the resulting nanocomposites with an intercalated or, more preferentially, exfoliated structure must exhibit strong interfacial interactions between the nanofiller surface and the polymer matrix. In our last paper, we have demonstrated that phosphonium ionic liquids (IL) can act as surfactants to LDH, which disperse the LDH nanoparticles in the PBAT matrix during the melt intercalation [8]. The improved compatibility between LDH and the PBAT matrix was shown to result in increased mechanical and water barrier properties of the PBAT/LDH nanocomposites as compared to neat PBAT. However, melt blending did not allow for full exfoliating of the LDH structure and thus only partially intercalated PBAT/LDH nanocomposites were prepared.

There exist several strategies in how to delaminate the multi-lamellar structure of LDH and to prepare fully exfoliated PBAT/LDH nanocomposites: Solvent intercalation, melt blending, and in-situ polymerization. During solvent intercalation, LDH is exfoliated using a solvent (e.g., chloroform) in which PBAT is soluble [15]. The method is not suitable for preparation of bulk PBAT/LDH nanocomposites since a volatile organic solvent is used and must be subsequently removed. The melt blending involves dispersing of LDH in PBAT in the molten state using a strong shear force [16,17]. This straightforward technique does not require any solvent and is most commonly used for the dispersion of cationic clays (smectitic clay minerals, e.g., montmorillonites, MMT) in PBAT [2], while its application for exfoliation of bulky anionic clays (LDH) was found more difficult due to their high charge density [18]. This method, however, does not permit full exfoliation of all LDH particles, and often a mixture of mono- and multi-lamellar sheets is received [8]. The in-situ polymerization consists of LDH swelling in the mixture of monomers, followed by polycondensation during which the chain growth in the interlayer space accelerates LDH delamination [19,20]. The exfoliation process of the swollen LDH layers is further supported by external forces via ultrasonic treatment or mechanical high shear mixing [18]. A suitable organic modification of LDH by anionic exchange is often necessary in order to facilitate monomer diffusion into the LDH interlayer space and at the same time initiate or catalyze the polymerization process. It has been shown that IL are ideal candidates for this kind of LDH treatment.

Ionic liquids (IL) are organic salts with low melting temperature (<100 °C) exhibiting several unique properties, e.g., excellent thermal and chemical stability, low vapor pressure and no flammability, which makes them suitable for instance as additives for polymers [21]. Moreover, ILs offer broad range of organic anions, which are not always commercially available as simple salts. Finally, the prices of ILs and simple salts are comparable. Therefore, their use for LDH (or in general inorganic filler) modification is convenient and furthermore mild conditions can be applied for filler modification. We have shown that organic anions of IL introduced into the LDH interlayer space are able to delaminate the LDH compounds and simultaneously initiate in-situ polymerization of ε-caprolactone [22].

In this paper, we use a similar approach to investigate in-situ prepared PBAT/LDH nanocomposites. First, Mg-Al LDH modified with different IL-anions are synthesized using direct co-precipitation. Contrary to the previously used anion exchange reaction producing mainly IL-surface-modified LDH [8], the co-precipitation technique enables it to reach higher content of intercalated IL-anions and thus to promote LDH exfoliation during polymerization. In the next step, PBAT nanocomposites are prepared in the presence of modified LDH via in-situ polycondensation. Finally, the influence of 1.5 and 5 wt % addition of IL-modified LDH on morphology, water/gas permeability, thermal, and mechanical properties of the PBAT/LDH nanocomposites is determined.

2. Materials and Methods

Trihexyltetradecylphosphonium bis (2,4,4-trimethylpentyl) phosphinate (IL-phosphinate), trihexyltetradecylphosphonium decanoate (IL-decanoate), and trihexyltetradecylphosphonium bis (2-ethylhexyl) phosphate (IL-phosphate) (see Table 1) were provided by IoLiTec (Heilbronn, Germany). $MgCl_2{\cdot}6H_2O$, $AlCl_3{\cdot}9H_2O$, NaOH, NH_4OH, NaCl, THF (p.a.), $CHCl_3$ (p.a.), ethanol (p.a.), and methanol (p.a.) were supplied by Lach-Ner (Neratovice, Czech Republic). All chemicals necessary for synthesis of LDH were used as received without further purification.

Table 1. Ionic liquids used for the modification of layered double hydroxides (LDH).

Ionic Liquid	Chemical Structure	Designation
Trihexyltetradecylphosphonium bis (2,4,4-trimethylpentyl) phosphinate	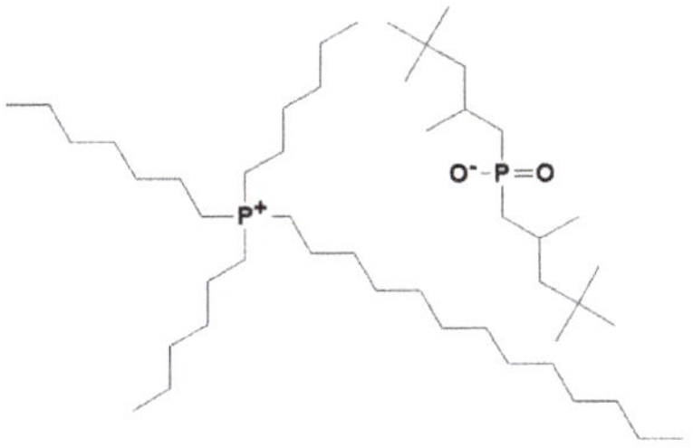 P^+ O^-–P=O	IL-phosphinate
Trihexyltetradecylphosphonium decanoate	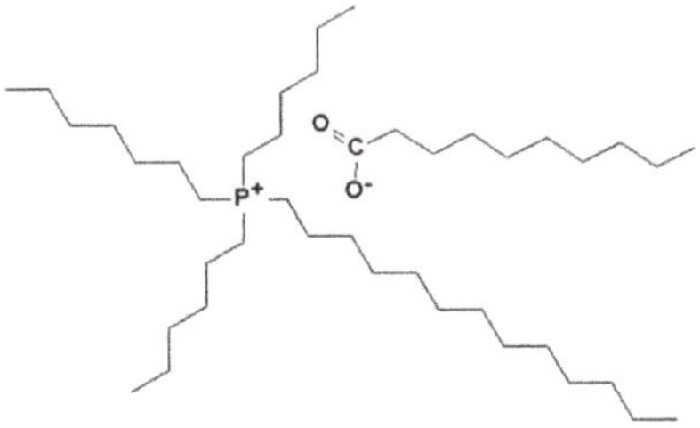 P^+ O=C–O^-	IL-decanoate
Trihexyltetradecylphosphonium bis (2-ethylhexyl) phosphate	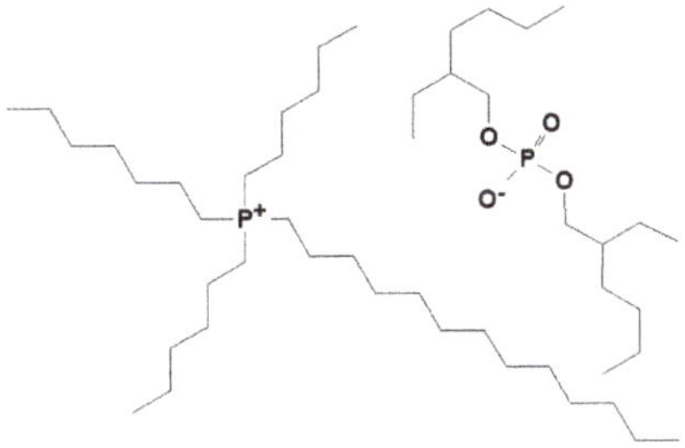 P^+ O, P, O, O, O^-	IL-phosphate

2.1. Preparation of LDH with Intercalated IL-Anions

Direct co-precipitation of metal salts in the presence of different ILs was used for the preparation of the LDH with intercalated IL-anions. Calculations were made for 2 g of LDH; ratio of IL/LDH was 1.5/1 (calculated according to anionic exchange capacity of LDH 3.35 meq/g). In a three necked flask, IL was first dissolved in a suitable organic solvent and subsequently an aqueous solution (25 mL) of $MgCl_2{\cdot}6H_2O$ and $AlCl_3{\cdot}9H_2O$ (Mg/Al molar ratio of 2/1) was added dropwise. The exact weights were as follows: 7.8 g (0.01 mol) of IL-phospinate was dissolved in 50 mL of THF and the solution of $MgCl_2{\cdot}6H_2O$ (4.3 g, 0.02 mol) and $AlCl_3{\cdot}9H_2O$ (2.6 g, 0.01 mol) in deionized water (25 mL) was added; 6.7 g (0.01 mol) of IL-decanoate was dissolved in 40 mL of $CHCl_3$ and the solution of $MgCl_2{\cdot}6H_2O$ (4.3 g, 0.02 mol) and $AlCl_3{\cdot}9H_2O$ (2.6 g, 0.01 mol) in deionized water (25 mL) was added; 8.1 g (0.01 mol) of IL-phosphate was dissolved in 40 mL of ethanol and the solution of $MgCl_2{\cdot}6H_2O$ (4.3 g, 0.02 mol) and $AlCl_3{\cdot}9H_2O$ (2.6 g, 0.01 mol) in deionized water (25 mL) was added. The molar IL/Al ratio of 1/1 was always used. In the case of IL-decanoate, we first tried to apply mixtures of miscible solvents but IL-decanoate remained dissolved, having no tendency to bond anions onto LDH and stack layers together. LDH with no significant IL-decanoate intercalation were obtained. Therefore, an immiscible $CHCl_3$/water mixture together with vigorous stirring led to swelling of the created LDH and its subsequent intercalation. The pH was kept constant at pH = 10 with the addition of NH_4OH. The resulting slurry was then aged at 60 °C for 24 h. Finally, the LDH precipitate was filtered, washed with a mixture of water and organic solvent, and dried at 80 °C for 48 h under vacuum. All steps were conducted under nitrogen atmosphere to avoid contamination by carbonate. Powders of three LDH modified with IL-phosphinate, IL-decanoate, and IL-phosphate were prepared and denoted as LDH-phosphinate, LDH-decanoate, and LDH-phosphate, respectively.

As reference material, Cl^--intercalated Mg-Al LDH (pristine LDH) was prepared by the co-precipitation method adapted from [23]. Briefly, 150 mL of an aqueous solution of NaOH (35 mmol) and NaCl (34 mmol) was slowly titrated with an ethanol-water solution of $MgCl_2{\cdot}6H_2O$ and $AlCl_3{\cdot}9H_2O$ (Mg/Al molar ratio of 2/1, cations concentration 0.375 mol/L) at 80 °C under nitrogen atmosphere. The formed precipitate was then aged for 1 h. The final sludge was filtrated on a Büchner funnel, washed with deionized water, and dried at 80 °C for 12 h to obtain the white powder of pristine LDH.

2.2. Preparation of PBAT/LDH Nanocomposites

Poly (butylene adipate-*co*-terephthalate) (PBAT)/LDH nanocomposites were prepared via an in-situ polycondensation of 1,4-butanediol (BD, >99%, Sigma-Aldrich, Saint Louis, MO, USA), dimethyl terephthalate (DMT, >99%, Sigma-Aldrich, Saint Louis, MO, USA), and dimethyl adipate (DMA, >99%, Sigma-Aldrich, Saint Louis, MO, USA) and catalyzed by 0.3% tetra-n-butyl orthotitanate (>97%, Sigma-Aldrich, Saint Louis, MO, USA) according to the two-step procedure adapted from [24]. Prior to the first step, a given amount of LDH powder was freshly dried (80 °C under vacuum) and then dispersed in BD using an ultrasound bath for 10 min. In the first step, BD (62 mmol, 20% molar excess) with dispersed LDH, DMT (21 mmol) and catalyst (0.15 mmol) were charged into a 100 mL three-neck flask equipped with a magnetic stirrer, a nitrogen inlet and a distillation column. The reaction took place under nitrogen flow at 190 °C for ca 1 h until methanol was completely distilled out. In the second step, DMA (31 mmol) was added into the reaction flask. As soon as methanol distillation was completed, the temperature of reaction mixture was increased to 220 °C and the pressure was gradually decreased under a final reduced pressure lower than 25 Pa for 2 h. The prepared highly viscous melt of PBAT was then cooled down to room temperature, dissolved in chloroform, precipitated into methanol, and dried at 60 °C under vacuum. The obtained nanocomposite was analyzed using ^1H NMR (Figure S1) and SEC (PS standards) showing the ratio of butylene adipate (BA)/butylene terephthalate (BT) units of 59/41 (theor. 60/40) and M_w of ca. 45,000 g/mol (M_w/M_n = 1.6). The LDH content in PBAT nanocomposites was 1.5 and 5.0 wt %. The same two-step procedure was used for the synthesis of reference neat PBAT.

For further characterization and testing, the PBAT and PBAT/LDH nanocomposites were prepared as films (the thickness of 200 μm) using compression molding (PTFE molds) at 130 °C with a compression force of 50 kN for 2 min.

2.3. Characterizations

Infrared (FTIR) spectra of the LDH samples were measured using the attenuated total reflectance (ATR) technique on a spectrometer Spectrum 100T FT-IR (PerkinElmer, Waltham, MA, USA) with a deuterated triglycine sulfate (DTGS) detector fitted with a Universal ATR accessory with a diamond/ZnSe crystal. All spectra were recorded in the wavenumber range of 650–4000 cm^{-1} at 16 scans per spectrum and 4 cm^{-1} resolution.

^1H NMR spectra (600 MHz) of the prepared PBAT and PBAT/LDH nanocomposites were obtained using a Bruker Avance III 600 MHz NMR spectrometer with $CDCl_3$ as the solvent at 25 °C. The chemical shifts are relative to TMS using hexamethyldisiloxane (HMDSO, 0.05 ppm from TMS) as the internal standard.

X-Ray diffraction (XRD) patterns were obtained using a high-resolution diffractometer explorer (GNR Analytical Instruments, Novara, Italy) equipped with a one-dimensional silicon strip detector Mythen 1K (Dectris, Baden, Switzerland). The CuKα radiation (wavelength λ = 1.54 Å) was produced by a sealed X-ray tube operated at 40 kV and 30 mA and monochromatized with Ni foil (β filter). The measurements were performed in Bragg-Brentano geometry in the range of 2θ = 2–70° with a step 0.2°. The exposure time at each step was 10 s.

Thermogravimetric analysis (TGA) of the LDH samples and the PBAT/LDH nanocomposite films was performed using a Pyris 1 TGA (PerkinElmer, Waltham, MA, USA) in a temperature range from 35 to 750 °C at a rate of 10 °C/min; the purge gas flow rate was fixed at 25 mL/min of nitrogen. Temperature of 5% weight loss ($T_{d5\%}$) was evaluated for the nanocomposite samples. Standard deviation of TGA measurement was under 5%.

Transmission electron microscopy (TEM) was performed on a Tecnai G2 Spirit Twin 12 microscope (FEI, BrNo, Czech Republic) in the bright field mode at the acceleration voltage of 120 kV. The PBAT/LDH nanocomposite films were cut into ultrathin sections (approximately 60 nm thick) by a Cryo-ultramicrotomy (Ultracut UCT, Leica, Wetzlar, Germany) using sample and knife temperatures of −80 °C and −40 °C, respectively.

Oxygen, carbon dioxide and water vapor transport properties of the PBAT/LDH nanocomposite films were examined by time-lag permeation method [25]. Each sample was inserted into a membrane cell which was then placed into a permeation apparatus and exposed to high vacuum (10^{-4} mbar) and a temperature of 45 °C for 12 h. After evacuation the temperature was set to 30 °C. Feed pressure p_i was 1.5 Bar. The permeability coefficient P was determined from the increase of the permeate pressure Δp_p per time interval Δt in a calibrated volume V_p of the product part during the steady state of permeation. For the calculation of the permeability coefficient, the following formula was used:

$$P = (\Delta p_p/\Delta t)\cdot[V_p l/(A p_i)]\cdot[1/(RT)] \quad (1)$$

where l is the membrane thickness, p_i feed pressure, A the area, T the temperature, and R the gas constant. Two or three specimens of each PBAT/LDH nanocomposite films were measured and the average reported. Relative standard deviations (SD) of Δp_p and Δt were lower than 0.3% (given by the 10 mbar MKS Barratron pressure transducer precision). Relative SD of membrane thickness measurement was 1%, relative SD of calibrated volume was lower than 0.5%, and relative SD of feed pressure was 0.3%. Therefore P values had the relative SD 2.4%. Gas diffusivities were estimated from the time-lag data, using the relation:

$$D = l^2/(6\theta) \quad (2)$$

where l is the film thickness and θ is the time lag. Relative SD of diffusion coefficients was 4%. Apparent solubility coefficients were calculated using the following equation:

$$S = P/D \tag{3}$$

The overall ideal selectivity (α_{ij}) of a polymer membrane for a pair of gases i and j is commonly expressed by the following relation:

$$\alpha_{ij} = P_i/P_j = (S_i/S_j)\cdot(D_i/D_j) \tag{4}$$

where P_i and P_j are pure gas permeabilities, D_i/D_j is the diffusion selectivity, and S_i/S_j is the solubility selectivity.

The thermal behaviors of the PBAT/LDH nanocomposite films were investigated using a DSC Q2000 calorimeter (TA Instruments, New Castle, DE, USA) with nitrogen purge gas (50 mL/min). The instrument was calibrated for temperature and heat flow using indium as a standard. Samples of about 10 mg were encapsulated into aluminum pans. Differential scanning calorimetry (DSC) was performed with a heating-cooling-heating cycle from −90 °C to 200 °C at 10 °C/min. Before and after the ramps, a two minute isothermal plateau was inserted. The glass transition temperature (T_g), melting temperature (T_m), and melting enthalpy (ΔH_m) were determined from the second heating scans. The crystallinity (X_c) of PBAT and PBAT nanocomposites was calculated using the following expression:

$$X_c = 100\cdot\Delta H_m/[\Delta H_m{}^o\cdot(1 - w_f)] \tag{5}$$

ΔH_m is the specific melting enthalpy of the sample, $\Delta H_m{}^o$ is the melting enthalpy of the 100% crystalline PBAT (114 J/g [26,27]), and w_f is the weight fraction of LDH filler.

The tensile tests on the PBAT/LDH nanocomposite films were conducted at ambient temperature using an Instron model 6025/5800R (Instron Limited, Norwood, MA, USA) equipped with a 100 N load cell at room temperature with a cross-head speed of 50 mm min^{-1}. Dumbbell-shaped specimens (ISO 527-3/5, half size) were used having the length of 60 mm, length and width of the narrow part: 16.5 and 3 mm, resp. and average thickness of ca 0.2 mm. Five specimens of each PBAT/LDH nanocomposite films were measured and the average reported.

3. Results and Discussion

3.1. *Synthesis of Layered Double Hydroxide with Intercalated Ionic Liquid Anions*

3.1.1. FTIR Spectra

The FTIR spectra of modified LDH (Figure 1) confirmed a significant content of organic phase due to the presence of IL-anions. A strong band of methylene deformation (PCH_2-) of phosphonium cations at 1466 cm^{-1}, typical for all the used phosphonium ILs (see FTIR spectra in the supplementary material, Figure S2) [28], was absent in all of the modified LDH (Figure 1b–d). The presence of phosphonium cations in all prepared IL-modified LDH was not detected. Therefore, the content of ILs adsorbed on the LDH surface can be neglected. It can be concluded that direct synthesis has led to a successful intercalation of IL-anions and the prepared LDH contains an organic part composed uniquely of the IL-anions.

The FTIR spectrum of LDH-decanoate (Figure 1c) exhibited strong asymmetric and symmetric carboxylate anion stretching bands at 1551 cm^{-1} and 1406 cm^{-1}, respectively [29,30], proving the presence of IL-decanoate anions in the LDH-decanoate.

The FTIR spectrum of LDH-phosphinate (Figure 1d) clearly evidences the presence of phosphinate anions as the band at 1467 cm^{-1} can be assigned to methylene deformation of PCH_2– [29] and the peaks at 1131 cm^{-1} and 1026 cm^{-1} correspond to asymmetric and symmetric (P = O) O-stretching [30].

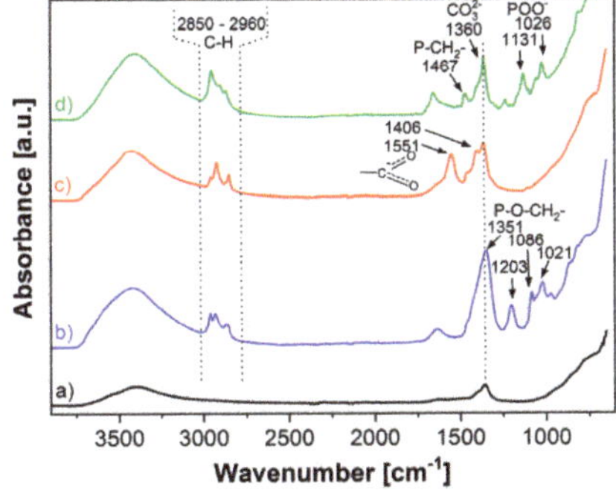

Figure 1. FTIR spectra of (**a**) pristine LDH, (**b**) LDH-phosphate, (**c**) LDH-decanoate, and (**d**) LDH-phosphinate.

The strong bands at 1351 cm^{-1} and the medium bands at 1203 cm^{-1}, 1086 cm^{-1}, and 1021 cm^{-1} in the FTIR spectrum of LDH-phosphate (Figure 1b) were assigned to P–O–CH_2– stretching [29], which evidences the presence of IL-phosphate anions.

Besides the IL-anions, the presence of carbonate anions in all of the modified LDH (as well as in the pristine LDH) was confirmed by the FTIR band at ca 1360 cm^{-1} (Figure 1) [31]. The direct synthesis was found to be highly efficient and led to preparation of LDH with a much higher content of organic anions (see the TGA results bellow) compared to the regeneration method published in our last study [22]. However, the presence of carbonate anions cannot be completely avoided, probably due to very high affinity of CO_3^{2-} to LDH [32]. The FTIR spectra of all LDH also contained absorption band at ca 1650 cm^{-1} typical for H–O–H deformation vibration of the interlayer water [11,33].

3.1.2. XRD Patterns

The XRD measurements enabled us to indicate whether the IL-anions were intercalated into the basal spacing of LDH. The XRD pattern of pristine LDH (Figure 2a) showed the formation of a single-phase, well-ordered crystalline-layered structure. The peaks located at $2\theta = 11.53°$, 23.38°, and 35.28° were attributed to the diffraction by (003), (006), and (009) planes, respectively. The (003) reflection of pristine LDH corresponded to the basal spacing value of 0.77 nm, which is typical for unmodified Mg-Al LDH [34–36].

The XRD patterns of the LDH modified with ILs demonstrated that the layered structure of the LDH was preserved during the one step LDH synthesis via co-precipitation. The presence of (003) reflection in the lower 2θ range—3.39°, 3.05°, and 2.19° for LDH-phosphate, LDH-decanoate, and LDH-phosphinate, respectively (Figure 2), denotes the basal spacing expansion to 2.6 nm, 2.9 nm, and 4.0 nm, respectively. This proves that the intercalation of IL anions, which size is larger than Cl^-, into the interlayer of LDH occurs. The obtained basal spacing expresses the sum of the thickness of one brucite-like octahedral layer and interlayer spacing. The later can be affected by the size and orientation of interlayer anion [36]. Assuming the thickness of brucite-like layer of ca 0.48 nm [37], then the interlayer spacing for LDH-phosphate, LDH-decanoate, and LDH-phosphinate expanded from 0.29 nm (for the pristine LDH) to approximately 2.1 nm, 2.4 nm, and 3.5 nm, respectively. The position and shape of broad XRD peaks at 10.77°, 11.20°, and 11.45° present for LDH-phosphate, LDH-decanoate, and LDH-phosphinate, respectively, suggest overlapping of two peaks. One of them could be ascribed to the shifted (006) reflection of intercalated LDH phase, while the second suggests possible generation of the second phase with smaller basal spacing. This is also supported by the shape of the peaks in $2\theta = 20°–23°$ present in all modified samples, which are also much broader than corresponding peaks in pristine LDH sample. The formation of the second phase could be caused by the different orientation of IL-anions inside the LDH interlayer spacing as was described by other authors [36] or by the generation of CO_3^{2-}-intercalated Mg-Al LDH (as supported by FTIR, Figure 1).

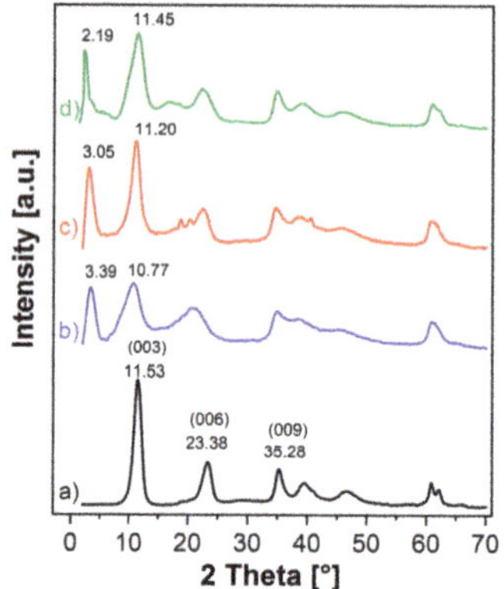

Figure 2. XRD patterns of (**a**) pristine LDH, (**b**) LDH-phosphate, (**c**) LDH-decanoate, and (**d**) LDH-phosphinate.

3.1.3. Thermogravimetric Analysis

Thermogravimetric (TG) and derivative weight TG curves of all LDH samples display an initial weight loss (8–13 wt %) between 50 °C and 250 °C (Figure 3) due to the release of physisorbed and interlayer water [17,22]. The TG curve of pristine LDH exhibits the removal of an interlayer carbonate anion and dehydroxylation of –OH groups between 250 °C and 500 °C (weight loss of ca 30 wt %) [38]. Moreover, TGA results of the modified LDH clearly show the degradation of intercalated IL-anions in the range of 250–500 °C. Significant weight loss of ca 42 wt % was observed in all modified LDH. Minimal contents of intercalated IL-anions in the modified LDH were estimated from the mass loss differences between the pristine LDH and the modified LDH to 11.8 wt % (LDH-phosphate), 12.6 wt % (LDH-decanoate), and 7.4 wt % (LDH-phosphinate, Figure 3). Unfortunately, the exact amount of intercalated IL-anions cannot be determined from TGA, because the IL-anion degradation, the interlayer CO_3^{2-} decomposition, and dehydroxylation of the metal hydroxides proceed simultaneously.

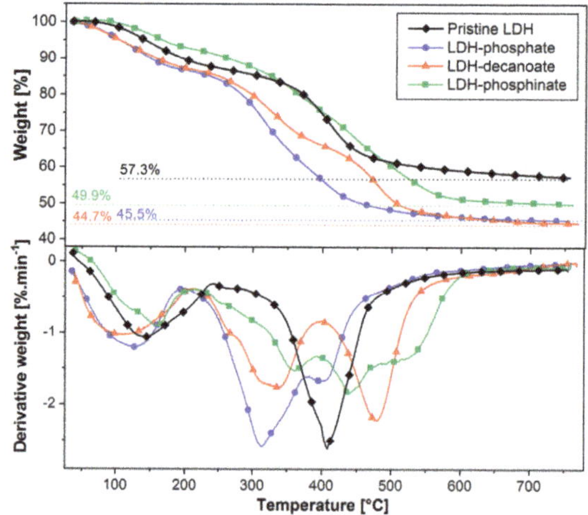

Figure 3. Thermogravimetric and derivative weight curves of pristine LDH, LDH-phosphate, LDH-decanoate, and LDH-phosphinate. The heating ramp was performed at 10 K·min^{-1} under N_2 atmosphere.

3.2. Characterization of PBAT/LDH Nanocomposites

3.2.1. Morphology

TEM images revealed that the type of LDH modification had strongly affected the final morphology of PBAT nanocomposites with 5 wt % of LDH filler (Figure 4). The non-modified LDH showed poor dispersion in the PBAT precursors with a low tendency to swell during the in-situ polymerization, which resulted in a composite containing large stacks of LDH (Figure 4a). Contrary to that, the successful IL-anion intercalation promoted LDH dispersion and swelling in the mixture of monomers resulting in the intercalated and exfoliated morphology of the final PBAT nanocomposites. Delamination of the modified LDH during the in-situ polymerization proceeded more easily due to weaker forces between organic IL-anions and the LDH layers as compared to small inorganic anions that hold the LDH layers together [39,40]. However, a fully exfoliated morphology of PBAT nanocomposites was not reached because a certain amount of CO_3^{2-}-intercalated into Mg-Al LDH was always present as a contaminant. The PBAT nanocomposites containing LDH-phosphate (Figure 4b) and LDH-phosphinate (Figure 4d) displayed single LDH layers (exfoliated structure) homogenously dispersed in the PBAT matrix although a few agglomerates of non-exfoliated LDH layers could also be observed, especially in the latter case. In contrast, the PBAT nanocomposite with LDH-decanoate (Figure 4c) showed the formation of LDH-rich domains in the PBAT matrix. The LDH-decanoate filler was swelled in monomers and delaminated during the in-situ polymerization (as evidenced from the XRD patterns—Figure 5) but the formed LDH sheets covered with IL-decanoate anions showed a low tendency to migrate into the surrounding PBAT matrix probably due to limited PBAT/IL-decanoate miscibility. It is known that a few % addition of phosphonium ILs into PBAT leads to a phase separation and formation of ionic clusters [41].

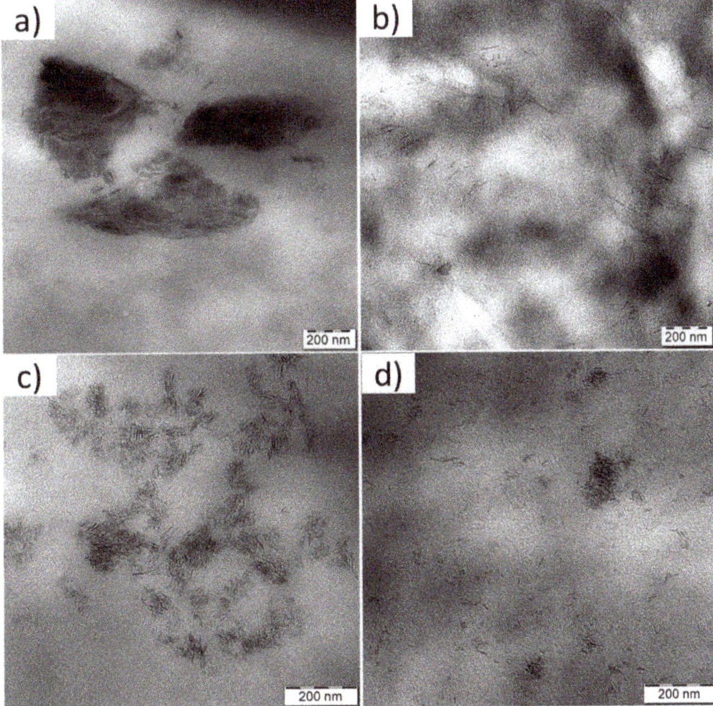

Figure 4. TEM images of PBAT nanocomposites containing 5 wt % of (**a**) pristine LDH, (**b**) LDH-phosphate, (**c**) LDH-decanoate, and (**d**) LDH-phosphinate.

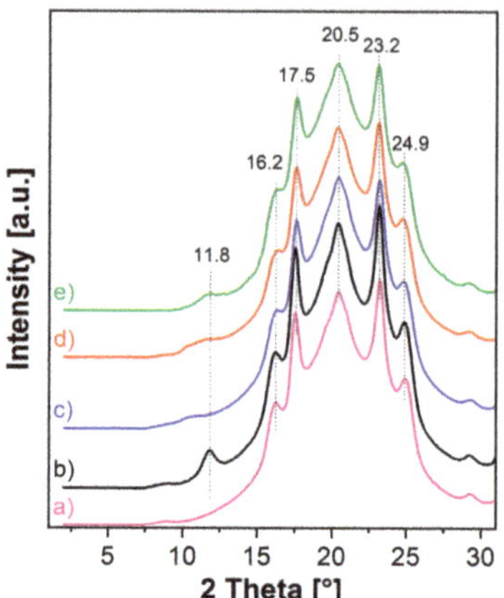

Figure 5. XRD patterns of (**a**) neat PBAT and PBAT nanocomposites with 5 wt % of (**b**) pristine LDH, (**c**) LDH-phosphate, (**d**) LDH-decanoate, and (**e**) LDH-phosphinate.

The XRD patterns of PBAT nanocomposites clearly confirmed the morphologies observed by TEM. The XRD pattern of PBAT nanocomposite with pristine LDH (Figure 5b) shows the intensive (003) diffraction peak at $2\theta = 11.8°$ (slightly shifted to higher angles due to a lower content of interlayer water in freshly dried LDH) revealing that the layered particles were neither exfoliated nor intercalated during the in-situ polycondensation. On the contrary, a complete disappearance of the reflection peak at a low angle range ($2\theta < 4°$) was observed for all PBAT nanocomposites with the modified LDH (Figure 5c–e) suggesting extensive exfoliation of the IL-anion-intercalated LDH particles. Moreover, in the cases of PBAT nanocomposites with the modified LDH, the XRD peak attributed to the second phase of LDH with smaller basal spacing, originally present at $2\theta = 10.77$–$11.45°$ (see Figure 2) was significantly decreased, broadened, and/or even shifted to lower angles (Figure 5c–e). It indicates that this LDH fraction was also partially delaminated and intercalated by PBAT chain during the in-situ polycondensation in the presence of IL-anion modified fillers.

The crystalline reflections associated with the PBAT matrix were observed at higher 2θ angles of 16.2°, 17.5°, 20.5°, 23.2°, and 24.9° (Figure 5) relating to the characteristic ($0\bar{1}1$), (010), ($\bar{1}\bar{1}1$), (100), and ($\bar{1}11$) planes, respectively [5,26]. The XRD data for all samples were identical in this region confirming that the LDH additions up to 5 wt % had not affected the crystalline structure of the PBAT matrix [26].

3.2.2. Water Vapor and Gas Barrier Properties

The transport properties of O_2, CO_2 gases, and water vapor in the neat PBAT and the PBAT nanocomposites were investigated and permeability (Table 2), diffusion (Table S1), and solubility (Table S2) coefficients, as well as their ideal selectivities were determined.

The permeability coefficients of all samples increase in the order of $O_2 < CO_2 << H_2O$ (Table 2). The permeability measurements were performed at temperatures at which the PBAT nanocomposites were in a rubbery state (above T_g—see the DSC result below). In such conditions, the permeation, diffusion, and solubility mechanisms in the system resembled gas transport in liquids. Crystallinity of the neat PBAT film (8%) was slightly enhanced by the presence of LDH fillers (<13%—see the DSC result below). However, such low overall crystallinity in the PBAT/LDH nanocomposites has a negligible effect on the evolution of gas transport properties [10].

Table 2. Comparison of O_2, CO_2, and water vapor permeabilities and ideal selectivities of neat PBAT and PBAT nanocomposites with 1.5 and 5 wt % of pristine LDH, LDH-phosphate, LDH-decanoate, and LDH-phosphinate.

	Permeability Coefficient (Barrer) [1]			Ideal Selectivity		
	O_2	CO_2	H_2O	CO_2/O_2	H_2O/O_2	H_2O/CO_2
Neat PBAT	1.55	16.9	2730	10.9	1770	162
+1.5% pristine LDH	1.42	15.5	2660	10.9	1880	172
+1.5% LDH-phosphate	1.35	14.3	1570	10.6	1170	110
+1.5% LDH-decanoate	1.48	16.1	2010	10.9	1358	125
+1.5% LDH-phosphinate	1.48	16.3	1980	11.0	1338	121
+5% pristine LDH	1.33	15.1	2550	11.4	1930	169
+5% LDH-phosphate	1.30	14.1	2300	10.8	1770	163
+5% LDH-decanoate	1.21	12.5	1530	10.3	1264	122
+5% LDH-phosphinate	1.43	14.0	1740	9.8	1217	124

[1] Barrer = 1×10^{-10} cm^3 (STP) $cm \cdot cm^{-2} \cdot s^{-1}$ $cmHg^{-1}$ = 3.3539×10^{-16} $mol \cdot s^{-1} \cdot m^{-1} \cdot Pa^{-1}$.

Since the permeability of gas and water vapor molecules was directly proportional to diffusion and solubility coefficients, the more dominating process can be determined based on the diffusion and solubility selectivities. In the cases of neat PBAT and PBAT/LDH nanocomposites, the gas and water vapor permeabilities were preferably driven by solubility in PBAT rather than by diffusion (the values of solubility selectivities were much higher than the diffusion ones) (Tables S1 and S2). Generally, solubility is correlated with the amount and intensity of interactions between the penetrant and polymer matrix [4,42,43].

It seems that the relatively high polarity of the PBAT backbone caused by the presence of ester groups promotes strong interactions between PBAT and polar molecules such as CO_2 and H_2O [2]. As a result, the determined solubility coefficients of CO_2 and H_2O are much higher than those of O_2, which further results in significantly faster permeation of these polar molecules through PBAT materials as compared to O_2. Neutral oxygen interacted weakly with PBAT and therefore relatively high selectivities for CO_2/O_2 gases were obtained.

The LDH fillers in PBAT nanocomposites were shown to slightly lower CO_2 and O_2 permeations by acting as a physical barrier, decreasing diffusion but only negligibly affecting solubility, which is in agreement with the theory of transport properties of polymeric materials filled with impermeable particles [44,45]. Gas molecules penetrate through free volume among the PBAT chains and the presence of LDH particles increases tortuosity of the gas molecule pathways. In our case, creation of longer diffusion paths for gas molecules depended mainly on the uniform dispersion and the aspect ratio of the 2D nanoplatelets. In the cases of nanocomposites with the modified LDH, the intercalated IL-anions promoted the LDH delamination, giving an abundance of homogenously dispersed LDH nanoplatelets throughout the PBAT matrix (Figure 4). This then contributed to the decreased values of diffusion coefficients at 5 wt % LDH (Table S1). Below this filler content, the effect of LDH addition and delamination was negligible due to a relatively low aspect ratio of the produced nanoplatelets. Unfortunately, the one-pot co-precipitation method of LDH synthesis in the presence of ILs did not allow for the formation of LDH with high lateral dimension as shown for materials synthesized by anion-exchange reaction, published in our previous study [8].

Contrary to CO_2 and O_2 transport properties, the neat PBAT exhibited high water vapor permeabilities (WVP, Table 2), which limits its use for food packaging [2,3]. The addition of pristine LDH resulted in a very slight WVP decrease (6.5%) due to poor filler dispersion in the PBAT matrix (Figure 4a). In contrast, the incorporation of modified LDH into the PBAT matrix led to significant WVP decreases varying in function of the chemical nature of the IL-anions. The most significant decrease in WVP coefficient was observed for the PBAT nanocomposite with 5.0 wt % of LDH-decanoate (44% reduction). However, it is of great importance to improve the water vapor barrier properties of PBAT at low nanofiller loadings to avoid viscosity increase and limited processability for food packaging films fabrication [3]. From the point of view of this application, the PBAT nanocomposite with 1.5 wt % of LDH-phosphate, in which WVP was reduced by 46% (Table 2) was selected as the most promising and used for comparison with other PBAT/filler nanocomposites with similar nanofiller loadings. Relative permeabilities (P_s/P_p), where P_s and P_p are WVP of PBAT/filler (in our case PBAT/LDH) nanocomposite and neat PBAT, respectively, are depicted in Figure 6.

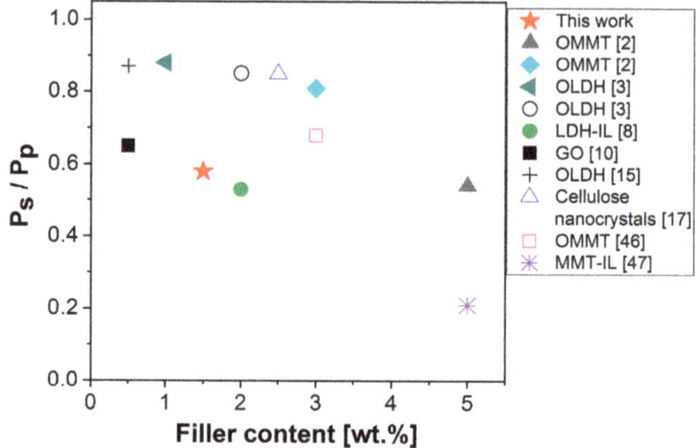

Figure 6. Comparison of relative permeability values (P_s/P_p) between this work (PBAT nanocomposite with 1.5 wt % of LDH-phosphate) and other PBAT/filler nanocomposites.

Our results are promising compared to the results so far published in the literature. They show the most significant improvement in the water vapor barrier properties of PBAT/LDH-phosphate nanocomposites, especially considering the relatively low nanofiller content (1.5 wt %). The materials prepared in our last study displayed similar WVP results only when higher amounts (2 wt %) of IL-modified LDH was incorporated into the PBAT matrix [8]. In that case, the PBAT nanocomposites exhibited partially exfoliated/intercalated morphology since the LDH delamination occurred to a minor extent due to the low content of intercalated IL. It seems that the vapor barrier properties of PBAT/LDH/IL systems are driven by the dispersion and exfoliation degree of LDH particles as well as by the formation of ionic clusters [8]. Highly polar surface (OH groups) of well dispersed LDH particles create more sorption sites (increased interactions) for polar water molecules which can more feasibly form clusters in the permeation pathways causing a significant decrease of water diffusion and thus overall decreased water permeation through the nanocomposites.

Other lamellar fillers (e.g., the most frequently used organically modified montmorillonite—OMMT, [2,3,46]) have been shown to provide similar WVP decrease of the PBAT nanocomposites at much higher OMMT loadings (e.g., 5 wt % [2], Figure 6). However, when non-polar phosphonium IL-modified montmorillonite is introduced, the relative permeability values of the PBAT/OMMT nanocomposites can be decreased even further (down to 0.2, Figure 6) [47].

3.2.3. Thermal Properties

Table 3 summarizes the results obtained by DSC. The glass transition temperature (T_g) values increased with the increasing content of the modified LDH containing IL-anions, which indicates homogenous nanofiller dispersion in the PBAT matrix and LDH delamination into individual LDH layers thus reducing mobility of PBAT chains. In contrast, the incorporation of pristine LDH showed no influence on T_g due to poor nanofiller dispersion in the PBAT matrix. The increase in crystallinity (X_c) and the decrease in melting temperature (T_m) of PBAT with increasing content of modified LDH indicates that the particles act as heterogeneous nucleating agents promoting the PBAT crystallite growth [8,9,15].

Table 3. DSC and TGA results of neat PBAT and PBAT nanocomposites with 1.5 and 5 wt % of pristine LDH, LDH-phosphate, LDH-decanoate, and LDH-phosphinate.

	T_g [°C]	T_m [°C]	ΔH_m [J/g]	X_c [%]	$T_{d5\%}$ [°C]
Neat PBAT	−40	116	9.7	8	354
+1.5% pristine LDH	−39	108	11.8	11	351
+1.5% LDH-phosphate	−36	111	10.1	9	336
+1.5% LDH-decanoate	−38	103	11.4	10	331
+1.5% LDH-phosphinate	−38	110	10.9	10	342
+5% pristine LDH	−40	113	8.3	8	346
+5% LDH-phosphate	−37	102	11.3	10	317
+5% LDH-decanoate	−37	99	13.6	13	310
+5% LDH-phosphinate	−35	101	13.2	12	324

The TGA results show that all prepared materials were thermally stable up to ca 280 °C. The LDH incorporation had slightly decreased the thermal stability ($T_{d5\%}$) of the PBAT matrix (Table 3). This effect was more significant for the modified LDH with IL-anions and thus probably connected to the degradation of IL-anions and possible catalytic effect of IL-anions on thermal degradation of PBAT at elevated temperature [15].

3.2.4. Mechanical Properties

Table 4 summarizes the results of uniaxial tensile properties of the prepared PBAT nanocomposite films. The LDH addition has led to a general improvement in stiffness (the increased Young moduli) as a consequence of increased rigidity induced by the nanofiller incorporation into the PBAT matrix and a slightly higher amount of the crystalline phase. However, tensile strength values have improved only in the PBAT/LDH nanocomposites with IL-phosphate and IL-phosphinate modifications. It means that these two modifications provided the most homogeneous LDH dispersion within the PBAT matrix. The best affinity between the LDH filler and PBAT was provided by IL-phosphinate as shown by the improved values of elongation at break. As we demonstrated in our last study, the IL-phosphinate modifier promotes formation of well-dispersed ionic clusters in PBAT, which can be responsible for increased ductility without reducing stiffness of the material [8,48]. In contrast, the decreased values of tensile strength and elongation at break of the PBAT nanocomposites containing LDH modified with IL-decanoate give evidence of material brittleness originating probably from the phase-separated morphology with LDH-rich domains in the PBAT matrix (see Figure 4). The brittle behavior was also observed for the PBAT nanocomposites with non-modified LDH as a result of poor nanofiller dispersion and a presence of large LDH agglomerates.

Table 4. Tensile properties of neat PBAT and PBAT nanocomposites with 1.5 and 5 wt % of pristine LDH, LDH-phosphate, LDH-decanoate, and LDH-phosphinate.

	Young Modulus [MPa]	Tensile Strength [MPa]	Elongation at Break [%]
Neat PBAT	76 ± 4	5.6 ± 0.3	122 ± 23
+1.5% pristine LDH	92 ± 4	5.9 ± 0.4	121 ± 18
+1.5% LDH-phosphate	110 ± 2	6.8 ± 0.4	133 ± 20
+1.5% LDH-decanoate	104 ± 4	5.0 ± 0.2	13 ± 3
+1.5% LDH-phosphinate	94 ± 2	8.7 ± 0.5	227 ± 15
+5% pristine LDH	99 ± 4	5.2 ± 0.5	29 ± 16
+5% LDH-phosphate	106 ± 4	6.2 ± 0.2	20 ± 3
+5% LDH-decanoate	137 ± 4	5.1 ± 0.1	10 ± 1
+5% LDH-phosphinate	112 ± 5	8.3 ± 0.6	225 ± 43

4. Conclusions

In this work, the preparation of PBAT/LDH nanocomposites by in-situ polycondensation in the presence of IL-anion modified LDH was reported.

First, a direct co-precipitation of metal salts in the presence of phosphonium ionic liquids (IL) was used for synthesis of IL-anion intercalated LDH. Using this technique, bis (2-ethylhexyl) phosphate (IL-phosphate), decanoate (IL-decanoate), and bis (2,4,4-trimethylpentyl) phosphinate (IL-phosphinate) anions were successfully introduced into the interlayer space of LDH compounds. The produced LDH show the increased interlayer spacing and high content of IL-anions. The results surprisingly show that LDH-IL prepared by this technique practically does not contain surface-bonded IL (in contrast to anion exchange technique used in our last study [8]). The way in how to prepare LDH containing both intercalated IL-anions and surface bonded IL is still challenging and under our investigation.

In the next step, LDH with intercalated IL-anions were shown to delaminate readily in a mixture of monomers during in-situ polycondensation. The produced PBAT/LDH nanocomposites exhibited exfoliated morphology either with homogenously dispersed LDH nanoplatelets in the PBAT matrix (the cases of IL-phosphate and IL-phosphinate intercalated LDH) or formation of LDH-rich domains in the PBAT matrix (the case of IL-decanoate intercalated LDH). Moreover, the IL-phosphinate modifier was shown to ensure the strongest LDH-PBAT affinity resulting in optimized mechanical performances. The presence of IL-anion intercalated LDH exhibited a relatively low effect on CO_2 and O_2 permeability reduction while the water vapor permeation was significantly decreased for all PBAT/IL-modified LDH nanocomposites. From this point of view, the produced PBAT/LDH nanocomposites are considered excellent candidates for food packaging applications.

Supplementary Materials: The following are available online at http://www.mdpi.com/2079-4991/9/4/618/s1, Figure S1: ^{1}H NMR spectrum of synthesized PBAT, Figure S2: FTIR spectra of (a) IL-phosphate, (b) IL-decanoate and (c) IL-phosphinate, Table S1: Dependence of oxygen, carbon dioxide and water vapor diffusion coefficient and ideal selectivity of neat PBAT and PBAT nanocomposites with 1.5 and 5 wt % of pristine LDH, LDH-phosphate, LDH-decanoate and LDH-phosphinate, Table S2: Dependence of oxygen, carbon dioxide and water vapor solubility coefficient and ideal selectivity of neat PBAT and PBAT nanocomposites with 1.5 and 5 wt % of pristine LDH, LDH-phosphate, LDH-decanoate and LDH-phosphinate.

Author Contributions: H.B., J.K. and S.L. conceived the idea and designed the experiments. J.P., S.B., E.P., J.H., S.A., M.K. and P.E. performed the experiments. H.B., J.K., J.P. and S.L. analyzed the data and wrote the paper. All the authors contributed to the preparation of the manuscript.

Funding: This research was funded by the Czech Science Foundation, grant number 17-08273S, and the Ministry of Education, Youth and Sports of CR, project POLYMAT LO1507.

Conflicts of Interest: The authors declare no conflict of interest.

References

1. Bordes, P.; Pollet, E.; Avérous, L. Nano-biocomposites: Biodegradable polyester/nanoclay systems. *Prog. Polym. Sci.* **2009**, *34*, 125–155. [CrossRef]
2. Li, J.; Lai, L.; Wu, L.; Severtson, S.J.; Wang, W.J. Enhancement of Water Vapor Barrier Properties of Biodegradable Poly (butylene adipate-co-terephthalate) Films with Highly Oriented Organomontmorillonite. *ACS Sustain. Chem. Eng.* **2018**, *6*, 6654–6662. [CrossRef]
3. Xie, J.; Wang, Z.; Zhao, Q.; Yang, Y.; Xu, J.; Waterhouse, G.I.; Zhang, K.; Li, S.; Jin, P.; Jin, G. Scale-Up Fabrication of Biodegradable Poly (butylene adipate-co-terephthalate)/Organophilic–clay Nanocomposite Films for Potential Packaging Applications. *ACS Omega* **2018**, *3*, 1187–1196. [CrossRef]
4. Someya, Y.; Sugahara, Y.; Shibata, M. Nanocomposites based on poly (butylene adipate-co-terephthalate) and montmorillonite. *J. Appl. Polym. Sci.* **2005**, *95*, 386–392. [CrossRef]
5. Fukushima, K.; Wu, M.H.; Bocchini, S.; Rasyida, A.; Yang, M.C. PBAT based nanocomposites for medical and industrial applications. *Mater. Sci. Eng.* **2012**, *32*, 1331–1351. [CrossRef]
6. Xu, G.; Qin, S.; Yu, J.; Huang, Y.; Zhang, M.; Ruan, W. Effect of migration of layered nanoparticles during melt blending on the phase morphology of poly (ethylene terephthalate)/polyamide 6/montmorillonite ternary nanocomposites. *RSC Adv.* **2015**, *5*, 29924–29930. [CrossRef]

7. Falcão, G.A.; Vitorino, M.B.; Almeida, T.G.; Bardi, M.A.; Carvalho, L.H.; Canedo, E.L. PBAT/organoclay composite films: Preparation and properties. *Polym. Bull.* **2017**, *74*, 4423–4436. [CrossRef]
8. Livi, S.; Lins, L.; Peter, J.; Benes, H.; Kredatusova, J.; Donato, R.; Pruvost, S. Ionic liquids as surfactants for layered double hydroxide fillers: Effect on the final properties of poly (butylene adipate-co-terephthalate). *Nanomaterials* **2017**, *7*, 297. [CrossRef]
9. Kashi, S.; Gupta, R.K.; Kao, N.; Hadigheh, S.A.; Bhattacharya, S.N. Influence of graphene nanoplatelet incorporation and dispersion state on thermal, mechanical and electrical properties of biodegradable matrices. *J. Mater. Sci. Technol.* **2018**, *34*, 1026–1034. [CrossRef]
10. Ren, P.G.; Liu, X.H.; Ren, F.; Zhong, G.J.; Ji, X.; Xu, L. Biodegradable graphene oxide nanosheets/poly-(butylene adipate-co-terephthalate) nanocomposite film with enhanced gas and water vapor barrier properties. *Polym. Test.* **2017**, *58*, 173–180. [CrossRef]
11. Guo, Y.; Wang, J.; Li, D.; Tang, P.; Leroux, F.; Feng, Y. Micrometer-sized dihydrogenphosphate-intercalated layered double hydroxides: Synthesis, selective infrared absorption properties, and applications as agricultural films. *Dalton Trans.* **2018**, *47*, 3144–3154. [CrossRef]
12. Qu, J.; Zhang, Q.; Li, X.; He, X.; Song, S. Mechanochemical approaches to synthesize layered double hydroxides: A review. *Appl. Clay Sci.* **2016**, *119*, 185–192. [CrossRef]
13. Theiss, F.L.; Ayoko, G.A.; Frost, R.L. Synthesis of layered double hydroxides containing Mg^{2+}, Zn^{2+}, Ca^{2+} and Al^{3+} layer cations by co-precipitation methods—A review. *Appl. Surf. Sci.* **2016**, *383*, 200–213. [CrossRef]
14. Kojima, Y.; Usuki, A.; Kawasumi, M.; Okada, A.; Fukushima, Y.; Kurauchi, T.; Kamigaito, O. Mechanical properties of nylon 6-clay hybrid. *J. Mater. Res.* **1993**, *8*, 1185–1189. [CrossRef]
15. Xie, J.; Zhang, K.; Wu, J.; Ren, G.; Chen, H.; Xu, J. Bio-nanocomposite films reinforced with organo-modified layered double hydroxides: Preparation, morphology and properties. *Appl. Clay Sci.* **2016**, *126*, 72–80. [CrossRef]
16. Hennous, M.; Derriche, Z.; Privas, E.; Navard, P.; Verney, V.; Leroux, F. Lignosulfonate interleaved layered double hydroxide: A novel green organoclay for bio-related polymer. *Appl. Clay Sci.* **2013**, *71*, 42–48. [CrossRef]
17. Morelli, C.L.; Belgacem, N.; Bretas, R.E.; Bras, J. Melt extruded nanocomposites of polybutylene adipate-co-terephthalate (PBAT) with phenylbutyl isocyanate modified cellulose nanocrystals. *J. Appl. Polym. Sci.* **2016**, *133*, 43678. [CrossRef]
18. Mao, N.; Zhou, C.H.; Tong, D.S.; Yu, W.H.; Lin, C.C. Exfoliation of layered double hydroxide solids into functional nanosheets. *Appl. Clay Sci.* **2017**, *144*, 60–78. [CrossRef]
19. Taviot-Guého, C.; Prévot, V.; Forano, C.; Renaudin, G.; Mousty, C.; Leroux, F. Tailoring hybrid layered double hydroxides for the development of innovative applications. *Adv. Funct. Mater.* **2018**, *28*, 1703868. [CrossRef]
20. Wang, G.A.; Wang, C.C.; Chen, C.Y. The disorderly exfoliated LDHs/PMMA nanocomposite synthesized by in situ bulk polymerization. *Polymer* **2005**, *46*, 5065–5074. [CrossRef]
21. Lu, J.; Yan, F.; Texter, J. Advanced applications of ionic liquids in polymer science. *Prog. Polym. Sci.* **2009**, *34*, 431–448. [CrossRef]
22. Kredatusová, J.; Beneš, H.; Livi, S.; Pop-Georgievski, O.; Ecorchard, P.; Abbrent, S.; Pavlova, E.; Bogdał, D. Influence of ionic liquid-modified LDH on microwave-assisted polymerization of ε-caprolactone. *Polymer* **2016**, *100*, 86–94. [CrossRef]
23. Milagres, J.L.; Bellato, C.R.; Vieira, R.S.; Ferreira, S.O.; Reis, C. Preparation and evaluation of the Ca-Al layered double hydroxide for removal of copper (II), nickel (II), zinc (II), chromium (VI) and phosphate from aqueous solutions. *J. Environ. Chem. Eng.* **2017**, *5*, 5469–5480. [CrossRef]
24. Perz, V.; Bleymaier, K.; Sinkel, C.; Kueper, U.; Bonnekessel, M.; Ribitsch, D.; Guebitz, G.M. Data on synthesis of oligomeric and polymeric poly (butylene adipate-co-butylene terephthalate) model substrates for the investigation of enzymatic hydrolysis. *Data Brief* **2016**, *7*, 291–298. [CrossRef]
25. Rutherford, S.W.; Do, D.D. Review of time lag permeation technique as a method for characterisation of porous media and membranes. *Adsorption* **1997**, *3*, 283–312. [CrossRef]
26. Chivrac, F.; Kadlecová, Z.; Pollet, E.; Avérous, L. Aromatic copolyester-based nano-biocomposites: Elaboration, structural characterization and properties. *J. Polym. Environ.* **2006**, *14*, 393–401. [CrossRef]
27. Herrera, R.; Franco, L.; Rodriquez-Galan, A.; Puiggali, J. Characterization and degradation behavior of poly(butyleneadipate-co-terephthalate)s. *J. Appl. Polym. Sci.* **2002**, *40*, 4141–4157. [CrossRef]
28. Witschard, G.; Griffin, C.E. Infrared absorption characteristics of alkyl and aryl substituted phosphonium salts. *Spectrochim. Acta* **1963**, *19*, 1905–1910. [CrossRef]

29. Silverstein, R.M.; Bassler, G.C. Spectrometric identification of organic compounds. *J. Chem. Educ.* **1962**, *39*, 546. [CrossRef]
30. Ha, J.U.; Xanthos, M. Novel modifiers for layered double hydroxides and their effects on the properties of polylactic acid composites. *Appl. Clay Sci.* **2010**, *47*, 303–310. [CrossRef]
31. Pizzoferrato, R.; Ciotta, E.; Ferrari, I.V.; Narducci, R.; Pasquini, L.; Varone, A.; Richetta, M.; Antonaroli, S.; Braglia, M.; Knauth, P.; et al. Layered double hydroxides containing an ionic liquid: Ionic conductivity and use in composite anion exchange membranes. *Chem. Electr. Chem.* **2018**, *5*, 2781–2788. [CrossRef]
32. Costa, F.R.; Leuteritz, A.; Wagenknecht, U.; Jehnichen, D.; Haeussler, L.; Heinrich, G. Intercalation of Mg–Al layered double hydroxide by anionic surfactants: Preparation and characterization. *Appl. Clay Sci.* **2008**, *38*, 153–164. [CrossRef]
33. Cavani, F.; Trifiro, F.; Vaccari, A. Hydrotalcite-type anionic clays: Preparation, properties and applications. *Catal. Today* **1991**, *11*, 173–301. [CrossRef]
34. Wang, D.Y.; Costa, F.R.; Vyalikh, A.; Leuteritz, A.; Scheler, U.; Jehnichen, D.; Wagenknecht, U.; Haussler, D.; Heinrich, G. One-step synthesis of organic LDH and its comparison with regeneration and anion exchange method. *Chem. Mater.* **2009**, *21*, 4490–4497. [CrossRef]
35. You, Y.; Vance, G.F.; Zhao, H. Selenium adsorption on Mg–Al and Zn–Al layered double hydroxide. *Appl. Clay Sci.* **2001**, *20*, 13–25. [CrossRef]
36. Kameda, T.; Saito, M.; Umetsu, Y. Preparation and characterisation of Mg–Al layered double hydroxides intercalated with 2-naphthalene sulphonate and 2,6-naphthalene disulphonate. *Mater. Trans.* **2006**, *47*, 923–930. [CrossRef]
37. Miyata, S. The Syntheses of Hydrotalcite-Like Compounds and Their Structures and Physico-Chemical Properties—I: The Systems Mg^{2+}-Al^{3+}-NO^{3-}, Mg^{2+}-Al^{3+}-Cl^{-}, Mg^{2+}-Al^{3+}-ClO^{4-}, Ni^{2+}-Al^{3+}-Cl^{-} and Zn^{2+}-Al^{3+}-Cl^{-}. *Clays. Clay Miner.* **1975**, *23*, 369–375. [CrossRef]
38. Soares, B.G.; Ferreira, S.C.; Livi, S. Modification of anionic and cationic clays by zwitterionic imidazolium ionic liquid and their effect on the epoxy-based nanocomposites. *Appl. Clay Sci.* **2017**, *135*, 347–354. [CrossRef]
39. Hou, X.; Bish, D.L.; Wang, S.L.; Johnston, C.T.; Kirkpatrick, R.J. Hydration, expansion, structure, and dynamics of layered double hydroxides. *Am. Min.* **2003**, *88*, 167–179. [CrossRef]
40. Li, Q.; Kirkpatrick, R.J. Organic anions in layered double hydroxides: An experimental investigation of citrate hydrotalcite. *Am. Miner.* **2007**, *92*, 397–402. [CrossRef]
41. Livi, S.; Bugatti, V.; Soares, B.G.; Duchet-Rumeau, J. Structuration of ionic liquids in a poly (butylene-adipate-co-terephthalate) matrix: Its influence on the water vapour permeability and mechanical properties. *Green Chem.* **2014**, *16*, 3758–3762. [CrossRef]
42. Giel, V.; Galajdová, B.; Popelková, D.; Kredatusová, J.; Trchová, M.; Pavlova, E.; Beneš, H.; Válek, R.; Peter, J. Gas transport properties of novel mixed matrix membranes made of titanate nanotubes and PBI or PPO. *Desalin. Water Treat* **2015**, *56*, 3285–3293. [CrossRef]
43. Poláková, L.; Sedláková, Z.; Ecorchard, P.; Pavlova, E.; Peter, J.; Paruzel, B.; Beneš, H. Poly (meth) acrylate nanocomposite membranes containing in situ exfoliated graphene platelets: Synthesis, characterization and gas barrier properties. *Eur. Polym. J.* **2017**, *94*, 431–445. [CrossRef]
44. Strawhecker, K.E.; Manias, E. Structure and properties of poly (vinyl alcohol)/Na^{+} montmorillonite nanocomposites. *Chem. Mater.* **2000**, *12*, 2943–2949. [CrossRef]
45. Bharadwaj, R.K. Modeling the barrier properties of polymer-layered silicate nanocomposites. *Macromolecules* **2001**, *34*, 9189–9192. [CrossRef]
46. Chen, J.H.; Yang, M.C. Preparation and characterization of nanocomposite of maleated poly (butylene adipate-co-terephthalate) with organoclay. *Mater. Sci. Eng.* **2015**, *46*, 301–308. [CrossRef]
47. Livi, S.; Sar, G.; Bugatti, V.; Espuche, E.; Duchet-Rumeau, J. Synthesis and physical properties of new layered silicates based on ionic liquids: Improvement of thermal stability, mechanical behaviour and water permeability of PBAT nanocomposites. *RSC Adv.* **2014**, *4*, 26452–26461. [CrossRef]
48. Livi, S.; Gérard, J.F.; Duchet-Rumeau, J. Ionic liquids: Structuration agents in a fluorinated matrix. *Chem. Commun.* **2011**, *47*, 3589–3591. [CrossRef]

Article

New Insights on the Fast Response of Poly(Ionic Liquid)s to Humidity: The Effect of Free-Ion Concentration

Jianxia Nie [†], Songhua Xiao [†], Rou Tan [†], Taihong Wang and Xiaochuan Duan *

Pen-Tung Sah Institute of Micro-Nano Science and Technology, Xiamen University, Xiamen 361005, China; 17859750523@163.com (J.N.); songhuaxiao@foxmail.com (S.X.); susan.austuin21@yahoo.com (R.T.); thwang@xmu.edu.cn (T.W.)

* Correspondence: xcduan@xmu.edu.cn

† These authors equally contributed to this work.

Received: 4 May 2019; Accepted: 13 May 2019; Published: 16 May 2019

Abstract: The swelling mechanism is widely used to explain the response of ionic liquids (ILs) or poly(ionic liquid)s (PILs) to moisture. While a fairly broad consensus has been attained, there are still some phenomena that are not well explained. As a complement to the swelling mechanism, we systematically studied the free volume theory in the rapid response and recovery of PIL humidity performance. We chose poly(1-ethyl-3-vinylimidazolium bromide) (PIL-Br), poly(1-ethyl-3-vinylimidazolium tetrafluoroborate) (PIL-BF_4) and poly(1-ethyl-3-vinylimidazolium bis(trifluoromethane sulfonimide)) (PIL-TFSI) as model materials and investigated the impact of PIL structure including anion type, film thickness and affinity to moisture on performance to obtain the humidity sensing mechanism for PILs based on free volume theory. Hence, we can combine free volume theory with the designed PIL structures and their affinity with moisture to obtain a high concentration of free ions in PIL sensing films. Furthermore, the PIL humidity sensors also show fast, substantial impedance changes with changing humidity for real-time monitoring of the human respiratory rate due to a fast response and recovery performance. Therefore, our findings develop a new perspective to understand the humidity performance of PILs based on free volume theory, resulting in fast response and recovery properties realized by the rational design of PIL sensing films.

Keywords: poly(ionic liquid)s; humidity sensing; free-ion concentration; fast response and recovery; respiratory rate monitoring

---

1. Introduction

The advent of air-stable ionic liquids (ILs) is a milestone in modern analytical science and ILs have been widely used as signal-enhancing elements in electroanalytical applications [1]. As is well known, ILs consist entirely of cations and anions with very specific properties [2], such as high ionic conductivity and negligible vapour pressure, and they have been used in many fields of electrochemical research [3]. In particular, the physical and chemical properties of ILs can be controlled to an unprecedented level through the rational design of cation–anion pairs. Thus, ILs are also known as designer materials or task-specific ILs [4]. Owing to these distinctive features, ILs could be designed and serve as the recognition elements for various sensing platforms [5]. Nevertheless, it should be noted that the direct usage of ILs as sensing materials in their liquid state is inconvenient [6]. It is necessary to immobilize ILs in solid devices for practical applications while keeping their attractive properties [7]. To address this issue, poly(ionic liquid)s (PILs), or polymerized ionic liquids [8], have been studied and represent a class of polymers (or polyelectrolytes) that feature an IL monomer in each repeating unit which is connected through a polymeric backbone to form a macromolecular architecture. Thus, PILs can alleviate the shortcomings, such as leakage and instability, of liquid electrolytes in electrochemical devices [9]. However, the IL moiety is covalently attached to the macromolecule in a PIL which means the organic cation or anion is restricted in mobility. Compared with ILs or IL/polymer mixtures where both the cations and anions are mobile, PILs are considered as single-ion conductors with relatively lower ionic conductivity [10]. Given that the ionic conductivity of PILs has a significant impact on their sensing performance, unfortunately there are relatively few reports on rationalizing the relationship between the PIL molecular structure and its ion transport properties. Therefore, it is truly a big challenge to obtain a fundamental understanding of a molecular-scale framework for designing PILs and its relationship with bulk properties.

PIL responses to chemical species include changes in optical, electric and charge transport properties, each of which have been explored for PIL chemical sensors [11–13]. In chemiresistive sensors, PILs are well suited materials as their adjustable conductivity permits a host of well-documented sensing mechanisms. For example, some of the earlier studies utilized PIL-based chemical sensors to detect analytes (such as moisture [14], CO_2 [15], volatile organic compounds (VOCs) [16], etc.) via a swelling mechanism. In this case, the PIL is well-doped and the analyte is absorbed causing a solvated IL moiety, reducing the viscosity of the PIL and increasing the ionic conductivity, thus creating a reduction in electrical resistance [13]. Under the guidance of this principle, Huang et al. [17] fabricated an optical humidity sensor using PIL photonic crystals as sensing materials. By rational combination of PIL and photonic structure, the optical humidity sensor can be used to rapidly, sensitively and visually detect environmental humidity via a colour change in the whole visible range [18]. To improve the conductivity of PIL-based sensors, single-walled carbon nanotubes [19] and $La_2O_2CO_3$ nanoparticles [20] were introduced with improved sensing properties because of the synergistic effect. Although the swelling mechanism [21] has been widely applied in IL-based sensors [14], there are still several points that need further improvements. For instance, according to the swelling mechanism, the IL-based humidity sensors often exhibit a fast response property, owing to a strong affinity with water molecules, which usually means that a long recovery time is required. However, this prediction conflicts with our previous experimental results that the PIL-based humidity sensor exhibits both a fast response and recovery properties. Hence, more research is needed for a comprehensive understanding of PIL-based sensors, especially in terms of ionic transport.

Owing to flexibility and easy processing, polymers are considered as attractive sensing materials for humidity sensors, and some polymeric humidity sensors are commercially available [22]. However, conventional polymeric sensing materials, including most neutral polymers or inorganic salt doped polymers, often display poor conductivity due to a low free-ion concentration, resulting in a slow response when exposed to humid conditions. A sub-minute or sub-second response can be achieved by rational design of PIL sensing materials with an adjustable free-ion concentration. Thus, on-line monitoring of humidity is expected to be realized and applied to various healthcare areas [23], such as pulmonary-function diagnostics [24], management of patients undergoing anaesthesia and critical care medicine [25]. In this regard, we systematically investigated the humidity sensing performance of imidazolium-based PIL films. We combine the free volume theory with the designed PIL structures including anion types, film thickness and affinity with moisture to obtain a high concentration of free-ions in PIL sensing films. Thus, the conductivity of PIL humidity sensors was found to vary sensitively over a wide range of relative humidity levels (RH) with a fast response and recovery properties. Furthermore, the PIL humidity sensors also confirmed a fast, substantial impedance change with a change in humidity for real-time monitoring of the human respiratory rate due to their fast response and recovery performance. Therefore, our findings develop a new perspective to understand the humidity performance of PILs based on free volume theory, resulting in a fast response and recovery properties realized by the rational design of PIL sensing films. It is highly expected that the PIL humidity sensors can be implemented for continuous monitoring of humidity in various arenas.

2. Materials and Methods

2.1. Chemicals and Reagents

1-vinyl-3-ethylimidazolium bromide (VEIm-Br) was obtained from Lanzhou Greenchem ILS, LICP, CAS, China. Then, 2,2′-azobis(2-methylpropionitrile) (AIBN), sodium tetrafluoroborate ($NaBF_4$) and lithium bis(trifluoromethane sulfonimide) (LiTFSI) were purchased from the Sigma-Aldrich Company (Shanghai, China) and were used as received without further purification. The water that was used was deionized.

2.2. Synthesis of Poly(1-ethyl-3-vinylimidazolium bromide) (PIL-Br)

In a typical synthesis process, 10.38 g of 1-vinyl-3-ethylimidazolium bromide, 30 mg of AIBN and 70 mL of absolute ethanol were added into a 250 mL round-bottom flask. The mixture solution was stirred for 2 h under nitrogen atmosphere and was then heated to 70 °C for 24 h. After that, the resulting PIL was washed with tetrahydrofuran (THF) and then dried at 70 °C overnight. The schematic diagram is as follows:

2.3. Synthesis of Poly(1-ethyl-3-vinylimidazolium tetrafluoroborate) (PIL-BF_4)

In this synthesis process, 2.04 g of PEVIm-Br was dissolved in a mixed solution composed of 9 mL of deionized water and 9 mL of absolute ethanol (solution A) under magnetic stirring in an ice-water bath for 30 min. Then, 1.10 g of $NaBF_4$ was dissolved in another mixed solution composed of 9 mL of deionized water and 9 mL of absolute ethanol (solution B) under magnetic stirring in an ice-water bath for 30 min. Subsequently, solution B was slowly added into solution A and then the mixture was stirred in an ice-water bath for 30 min and was left to stand for 1 h. Finally, the resulting yellowish white solid was cleaned with deionized water and absolute ethanol several times before drying at 70 °C for 24 h. The schematic diagram is as follows:

2.4. Synthesis of Poly(1-ethyl-3-vinylimidazolium bis(trifluoromethane sulfonimide)) (PIL-TFSI)

The protocol for the synthesis of PIL-TFSI is similar to that of PIL-BF_4: 2.04 g of PIL-Br reacts with 2.81 g of bis(trifluoromethane sulfonimide) through an ion exchange reaction and the resulting yellowish solid was finally gained. The schematic diagram is as follows:

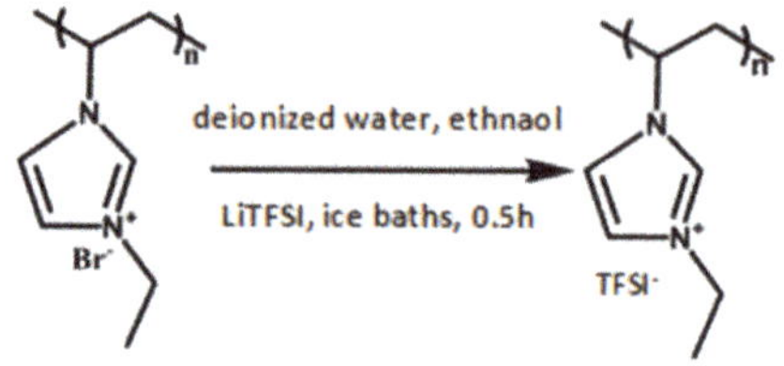

2.5. Characterization

The as-prepared PILs were characterized by Fourier-transform infrared spectroscopy (FTIR), ^{1}H NMR spectra, etc. TMDSC (NETZSCH DSC 204 F1) was used to determine the T_g of PIL-Br equilibrated at varying RH levels under nitrogen. First, a 250 mL bottle containing saturated aqueous solutions of K_2SO_4 salts was used to equilibrate PIL-Br for 2 h. Then, the sample was placed into a non-hermetically sealed aluminium differential scanning calorimeter (DSC) pan for measurement. The TMDSC runs consisted of cooling at a rate of 5 °C/min to −60 °C, followed by heating to 100 °C (with the same heating rate of 5 °C/min) and a temperature modulation of ±0.80 °C every 60 s. The DSC heats from 30 to 250 °C at a heating rate of 10 °C/min.

2.6. Humidity Sensor Preparation and Measurements

The humidity sensors were fabricated by spin-coating the PIL dispersion onto pre-cleaned interdigitated electrodes. Alumina ceramic was used as the substrate due to its high dielectric strength and excellent stability. Interdigitated electrodes composed of 10 nm of titanium and 80 nm of gold with a linewidth of 80 µm were fabricated by sputtering, as shown in Figure S1.

Spin-coating PIL onto the interdigitated electrodes: 0.06 g of PIL-Br was added into 600 mL of absolute ethanol and the mixture solution was then stirred for 2 h until completely dissolved. The sensors were fabricated by spin coating 4.0 µL of the mixture solution onto the pre-cleaned interdigitated electrodes at a speed of 4000 rad/s for 60 s. After that, the sensor was dried at room temperature for 48 h. For the case of PIL-BF_4, the solution was replaced by *N,N*-dimethylformamide (DMF) with the other conditions preserved to be the same. The PIL-Br thickness can be easily tuned by adjusting the solution concentration and revolving speed.

Humidity sensing properties were tested by an impedance analyser (Agilent 4294, 40 Hz to 110 MHz). The applied voltage amplitude was 500 mV and the measuring temperature was approximately 25 °C. Several 250 mL bottles containing different saturated salt solutions including LiCl, $MgCl_2$, K_2CO_3, NaBr, NaCl, KCl and K_2SO_4 were used as the sources of humidity for 11%, 33%, 43%, 59%, 75%, 85% and 98% RH, respectively.

3. Results and Discussion

3.1. Structural Characterization of PILs

The obtained PILs were first characterized by FTIR and ^{1}H NMR. The FT-IR spectra for the IL and PILs are shown in Figure 1 and the absorption peaks for the imidazolium rings in the IL and PILs are obvious. The C–H bonds in the imidazolium rings show a deformation vibration in-plane peak at 1170 cm^{-1} and external bending vibration peaks at ~750 cm^{-1}, ~650 cm^{-1}. The absorption peak at ~3140 cm^{-1} arises from the stretching vibration of the C–H bonds, and that at ~1570 cm^{-1} is due to the C=N bonds in the imidazolium rings. Thus, the results from the FT-IR spectra can clearly confirm the integrity of the imidazolium ring in the PILs. However, a complete disappearance of the band corresponding to Br^- (960 cm^{-1}) is observed with the appearance of new bands attributed to BF_4^- (1053 cm^{-1}) or TFSI$^-$ (~1187 cm^{-1}), which also proves the success of the anion exchange. In addition, the absorption peak at 3450 cm^{-1} is attributed to the hydroxyl groups (H–O) of water molecules mechanically adsorbed by the IL or PILs.

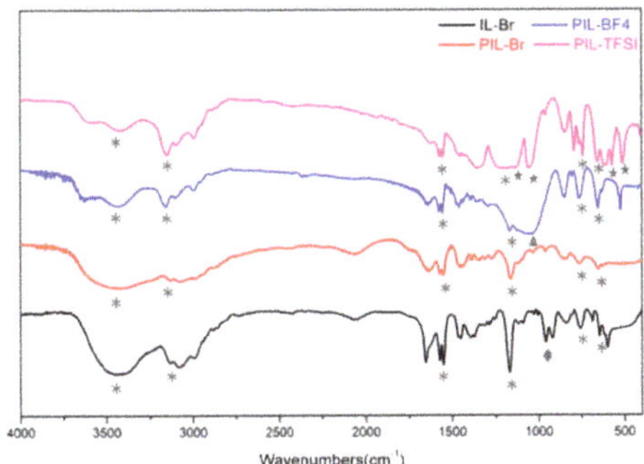

Figure 1. FT-IR spectra for IL-Br, PIL-Br, PIL-BF_4 and PIL-TFSI.

The ^{1}H NMR spectra for PIL-Br and IL-Br is shown in Figure 2 and the chemical formulae are shown in the illustration. The vinyl signals at 7.5, 6.1 and 5.4 ppm are observed to disappear, with the appearance of new signals at 2.7 and 4.5 ppm attributed to the protons of the polymeric backbone. Under the influence of the chemical group, the transformation and shift of signals near 10.5 and 8 ppm are distinct. We ensure that the polymerization by IL-Br monomer is successful via ^{1}H NMR test.

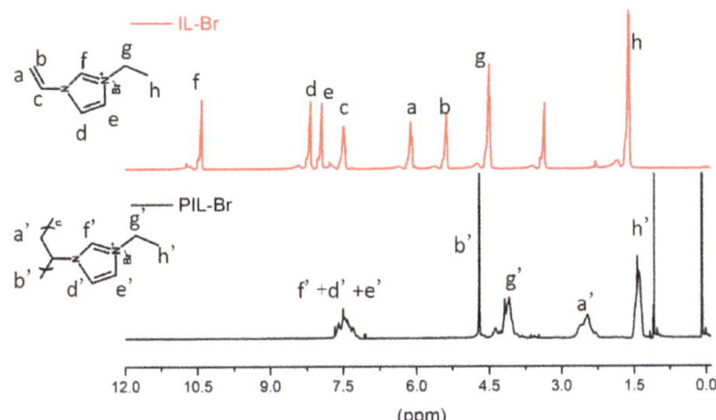

Figure 2. 600-MHz ^{1}H NMR spectra for PIL-Br and IL-Br.

Wetting, usually associated with hydrophilicity, is a phenomenon based on two-phase interface changes from a solid-gas interface to a solid-liquid interface. The wettability of a film material can be expressed by the contact angle (1–180°) and a large contact angle indicates a worse wettability. The static water contact angles for PIL-Br (34.6°), PIL-BF_4 (52.9°) and PIL-TFSI (97.8°) are shown in Figure 3, revealing the hydrophilicity: PIL-Br>PIL-BF_4>PIL-TFSI. The result corresponds to the effect of the anionic polarity to water molecules in theory.

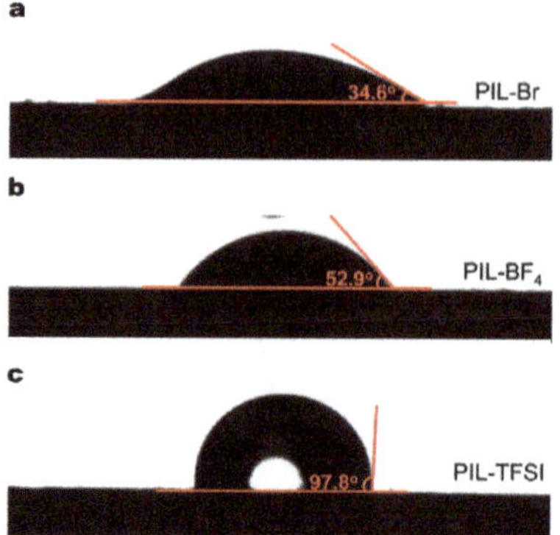

Figure 3. Static water contact angles corresponding to the as-prepared poly(ionic liquid)s (PILs): (**a**) PIL-Br, (**b**) PIL-BF_4 and (**c**) PIL-TFSI.

3.2. Swelling Mechanism for a Hydrophilic PIL-Br Humidity Sensor: The Effect of Thickness

When polymers are used in humidity sensing, the swelling process is usually exploited: (i) diffusion of water molecules through the polymer matrix, (ii) relaxation of polymer chains and (iii) expansion of the polymer network upon relaxation. To confirm that the humidity performance of the PILs can be affected by their swelling behaviour, a series of controllable experiments were exemplified using hydrophilic PIL-Br. Since the swelling behaviour of a PIL sensing film is directly related to its thickness, we thus investigate the effect of PIL-Br thickness on the humidity performance. The thickness of PIL-Br sensing films can be easily controlled by adjusting the concentration of the spin-coating solution, denoted as PIL-Br-10, PIL-Br-20 and PIL-Br-30, respectively. It is well known that the working frequency has an important impact on the humidity performance of a PIL sensor. To explore the optimum frequency for the PIL humidity sensors, we first investigated the impedance as a function of RH at different frequencies. The relationships between the impedance modulus of PIL-Br-10 and RH with measuring frequency from 100 Hz to 25 kHz are shown in Figure 4a. In the low frequency region from 100 Hz to 1 kHz, the PIL-Br-10 sensor presents good linearity with a total impedance modulus change of more than two orders of magnitude, which indicates high sensitivity in the whole RH range from 11% to 98% RH. However, the impedance curve gradually turns flat and the total impedance modulus change is reduced when the measuring frequency is higher than 2.5 kHz. This is mainly because the polarization of the water molecules cannot catch up with the directional change of the electrical field under high frequency and only electronic-displacement polarization and ion-displacement polarization can take place. Since the capacitance and dielectric constant of the PIL sensor is very small, it becomes independent of RH at high frequencies. The best linearity of the impedance versus RH curve for the PIL-Br-10 sensor appears at 500 Hz; thus, 500 Hz was defined as the ideal working frequency for the PIL-Br-10 sensor and was used in the following measurements. Similarly, the optimum measuring frequencies for PIL-Br-20 and PIL-Br-30 can be calculated to be 500 Hz (Figure 4d) and 2.5 kHz (Figure 4g), respectively. The dynamic impedance response of PIL-Br sensors within the step changes from 11% to 98% RH followed by a return to the initial state were also performed, as shown in Figure 4b,e,h. All sensors exhibited a good reversibility during the step change in the humidity levels, revealing good reliability of the as-prepared humidity sensors. In addition, humidity hysteresis—another criterion that is commonly used to estimate the reliability of humidity sensors—was also tested for PIL-Br sensors with varying thickness. The humidity hysteresis is defined as the maximum difference of the humidity sensor between the adsorption and desorption process [26]; thus, the humidity hysteresis was calculated to be 3.2% RH (Figure 4c), 4.3% RH (Figure 4f) and 5.2% RH (Figure 4i) for PIL-Br-10, PIL-Br-20 and PIL-Br-30 within the working range of 11%–98% RH, respectively. This result can further confirm the good reliability of PIL-Br humidity sensors.

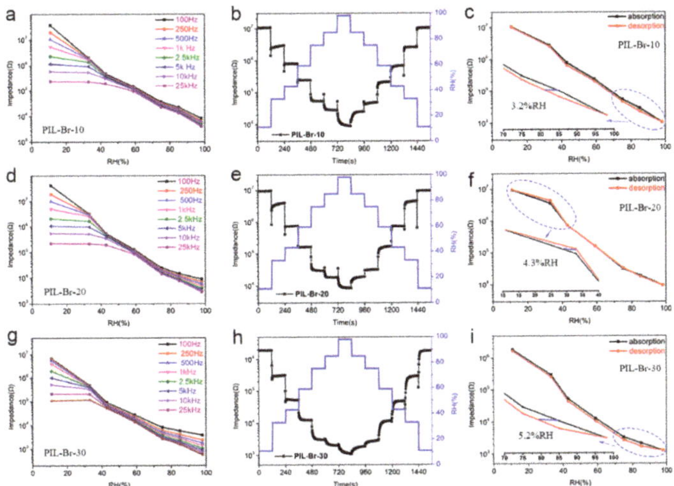

Figure 4. Thickness-dependent performance of PIL-Br humidity sensors: the impedance at varying RH levels measured under various frequencies: (**a**) PIL-Br-10, (**d**) PIL-Br-20, (**g**) PIL-Br-30, respectively; the dynamic response of PIL-Br sensors for increasing RH from 11% to 98% RH followed by a return to the initial state: (**b**) PIL-Br-10, (**e**) PIL-Br-20, (**h**) PIL-Br-30, respectively; the humidity hysteresis of (**c**) PIL-Br-10, (**f**) PIL-Br-20 and (**i**) PIL-Br-30, respectively.

The repeatability of PIL-Br sensors was further characterized by switching the measurement environments between 11% RH and various RH for five cycles, as shown in Figure 5a–c. All the PIL-Br sensors with varying thicknesses possess good repeatability. At the same time, the response and recovery times of these sensors were characterized in Figure 6. The sensor with the thinnest film has pretty short response and recovery times (6 s/10 s). In addition, we wondered the differences between these PIL-Br sensors in long-term stability, so we tested these sensors repeatedly per 15 days under the fixed humidity levels. The PIL-Br-10 sensor can maintain quite a stable dynamic response, as shown in Figure 5d. The data show good consistency of the PIL-Br-10 sensor at each humidity region. These results demonstrate that the as-prepared PIL-Br-10 sensors possess high stability and durability in a wide RH range. In contrast, the long-term stabilities of two other sensors which have thicker film are not good enough. This can be seen in Figure S2.

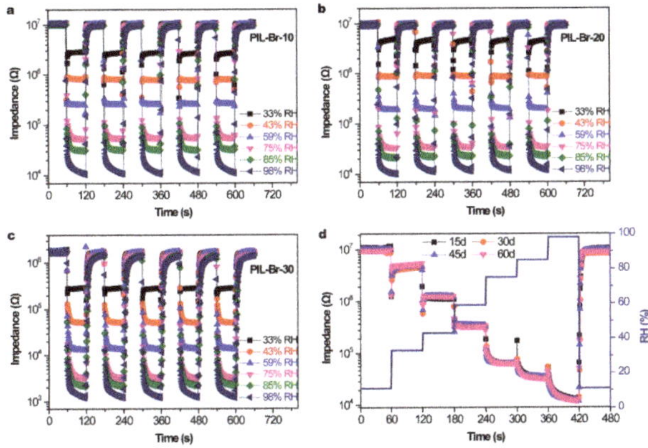

Figure 5. The repeatability characteristic of PIL-Br sensors from 11% to 98% RH: (**a**) PIL-Br-10, (**b**) PIL-Br-20 and (**c**) PIL-Br-30, respectively, (**d**) Long-term stability of a PIL-Br-10 humidity sensor.

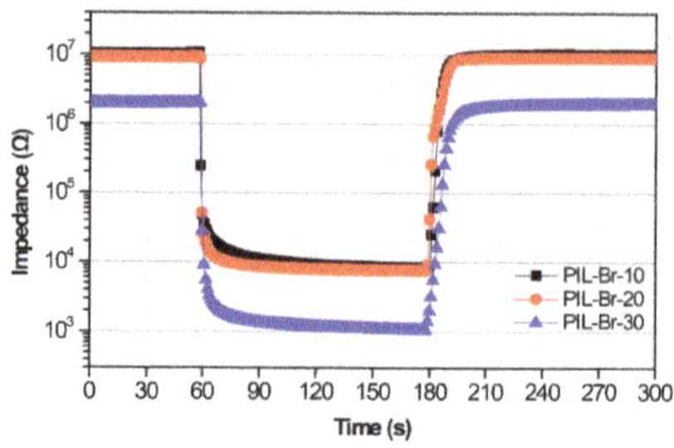

Figure 6. Response and recovery properties for PIL-Br humidity sensors with varying thickness.

For further discussion, the experimental impedance results obtained for the PIL-Br humidity sensors were fitted via equivalent circuits by using Z-View software, as shown in Figure 7. Here, R_S is related to the resistance of the electrode/film, R_1 is the bulk resistance of film, Z_{CPE} is a constant phase element (CPE) and Z_W is a Warburg impedance [27]. As shown in Figure 7a, the total impedance of the PIL-Br humidity sensors almost achieves a three orders of magnitude response and superior linear correlation coefficient under RH ranging from 11%–98%, which is advantageous for humidity sensors. The values for R_S and Z_{CPE} were hardly affected by ambient humidity, as shown in Figure 7b,d, because R_S relates only to the inter-contact between the electrode and film and Z_{CPE} is characterized as the influence of polarization in the sensing film. In addition, R_W decreases drastically as RH increases from 43% to 98%, leading to high-concentration and fast-action of mobile ions in the diffusing process due to the formation of continuous multi-water layers at 33%RH(PIL-Br-30) or 44%RH(PIL-Br-10, PIL-Br-20). It is worth noting that continuous multi-water layers can promote jumping of hydronium ions (H_3O^+) and free hydrogen ions (H^+) for conduction based on the Grotthuss mechanism [28]. Hydrogen-ion jump conduction occurs when a H^+ (H_3O^+) passes from one end to the other through an adjacent water molecule, rather than a H^+ (H_3O^+) going from one end to the other, with the former being faster. The value of R_1 decreases linearly by approximately three orders of magnitude within the full measured RH range (Figure 7c), which can be attributed to the improved conductance of the PIL-Br sensing film because of the increased number of absorbed water molecules. In addition, the results show that it is easy to increase the conductivity in thicker films because of the absorption of more water molecules. In summary, the PIL-Br-30 sensor has pretty outstanding conductivity during the ion diffusion process.

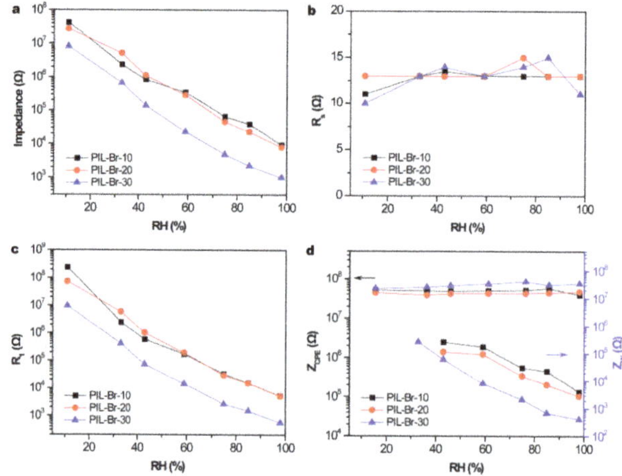

Figure 7. Dependence of the values of (**a**) total, (**b**) Rs, (**c**) R_1 and (**d**) Z_{CPE} and Z_W on varying RH levels for PIL-Br sensors with different thicknesses.

3.3. Humidity Mechanism for PILs Based on Free Volume

As mentioned above, PILs are usually considered to be single-ion conductors that are comprised of polymeric imidazolium-based cations and anions. In these cases, cations are structurally constrained as part of the polymer skeleton, which largely restricts their movement. Owing to the asymmetric volume of the cation and anion, there are large spaces (i.e., holes and voids) in the PILs. These spaces are also denoted as free volume, which can provide sorption sites to accommodate penetrate of water molecules and to allow ion-movement through the PIL matrix [29]. The permeation of water molecules through the free volume can induce a swelling of the PILs, thus altering their properties. Kasapis et al. [30] reported the swelling behaviour of polymers exposed to a hot and humid environment. The hygroscopic swelling of a polymer can cause a change in volume, similar to the thermal expansion caused by an increase in temperature, leading to a linear expansion of the free volume (f) of the polymer. Following derivation from the Williams-Landel-Ferry (WLF) expression [31], the time-temperature-humidity equivalence principle can be expressed as follows:

$$f = f_0 + e_T(T - T_0) + e_M(M - M_0) \quad (1)$$

where f is the free volume of a polymer, f_0 is the free volume fraction of a sample at the reference temperature T_0; e_T and e_M is the free volume fraction of the thermal expansion coefficient and humidity swelling coefficient, respectively; M_0 is the water content of a sample at the reference temperature T_0. Assuming that a sample remains in the same ambient temperature, the free volume will increase as the ambient humidity increases. According to the free volume theory, the relationship between the viscosity (η) and the free volume (f) of the polymer follows the Doolittle equation [32]:

$$\ln \eta = \ln A + B\left[\frac{1}{f} - 1\right] \quad (2)$$

where A and B are constants for a given material. Thus, the viscosity of PILs will decrease as the free volume increases. Furthermore, the relationship between conductivity (η) and viscosity (σ) of the PILs can be built by the following Walden equation:

$$\lg\sigma = \lg k + \alpha \lg\eta^{-1} \quad (3)$$

where k is a temperature dependent constant and α is a fitting parameter [33]. Thereby, the conductivity in PILs is inversely proportional to the viscosity and proportional to the free volume. Therefore, the conductivity of the PIL sensing films can be optimized by adjusting their free volume, which is conducive to the design of sensing materials for better performance.

Generally, the glass transition temperature (T_g) is defined as the temperature at which the mechanical properties of a polymer material radically change due to internal movement of the polymer chains [34]. For the as-prepared PILs, a lower T_g promotes a higher ionic conductivity. To verify that the PIL conductivity changes with a change in humidity based on free volume theory, we take PIL-Br as an example and investigate its T_g change under varying RH levels using temperature-modulated differential scanning calorimetry (TMDSC). From Figure 8a, the T_g of PIL-Br under dry conditions was measured to be 185.2 °C according to a DSC scan. For comparison, a typical TMDSC scan obtained from PIL-Br equilibrated to 98%, as shown in Figure 8b. The endotherm of the total signal at approximately 88 °C, which corresponds to the evaporation of the absorbed water, is exclusively present in the non-reversing heat-flow signal. The inflection point in the reversing signal at 40.3 °C occurs at a temperature that is well below the onset of water evaporation and designates the T_g of PIL-Br under a 98% RH condition. Thus, the T_g of PIL-Br decreases with increasing RH, which benefits improved ion movement. According to the Vogel-Tamman-Fulcher (VTF) equation: $\sigma = \sigma_0 e^{-B/T-T_0}$, where σ is the conductivity, σ_0 and B are constants and T_0 is the Vogel temperature where ion motion first occurs

($T_0 = T_g - 50\ K$) [35]. Therefore, the conductivity of PIL-Br increases under a high RH condition, which is well consistent with our hypothesis.

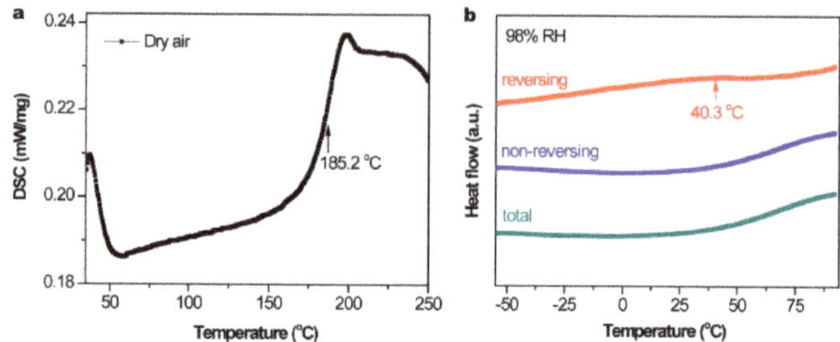

Figure 8. (**a**) DSC thermogram of PIL-Br equilibrated under dry conditions. (**b**) TMDSC thermogram of PIL-Br equilibrated under 98% relative humidity levels (RH): total, reversing and non-reversing heat-flow signals.

3.4. The Effect of Anion Type on Humidity Sensing Performance of PIL

Molecules with polar groups, which have a great affinity with water, can attract water molecules or be dissolved in water easily. Furthermore, the excellent hydrophilic PILs can absorb a large volume of water to increase their free volume according to the WLH equation. Thus, it is interesting to study the sensing performance of different hydrophilic PILs films. Adjusting the anion type in PILs film is an effective route to change free volume. A series of controllable experiments were designed using PILs with different anions to investigate the effect of anion type on humidity performance. PIL-BF_4 and PIL-TFSI were applied as humidity materials in our work because of their different hydrophilic performances. The method used to spin-coat PIL onto the interdigitated electrodes is similar to that used for the humidity sensor based on PIL-Br, as shown in Supporting S1. The humidity performance of the PIL sensors was tested at the working frequency, as shown in Figures S3–S8. To clearly observe the effect of anion type, we list the performance parameters for the PIL-based humidity sensors in Table 1. Here, *S/T-res* is defined as the ratio of the sensitivity to response time and, similarly, *S/T-rec* is defined as the ratio of the sensitivity to recovery time. These ratios are perfect for reflecting the velocity for response or recovery, especially for a response with different orders of magnitude. Figure 9 shows the performance of the PIL-based humidity sensors, with data far away from the origin of the coordinates representing high sensitivity, high response and recovery velocity. The different level of performance among the three PIL-based humidity sensors could be clearly distinguished, with the PIL-Br humidity sensors showing an obviously higher sensitivity, response and recovery velocity.

Table 1. The performance parameters of the as-prepared PIL-based humidity sensors.

Sensor	Sensitivity	Response Time (s)	Recovery Time (s)	S/T-res	S/T-rec
PIL-Br-10	1190	6	10	198.3	119.0
PIL-Br-20	1392	7	11	198.9	126.5
PIL-Br-30	2916	11	11	265.1	265.1
PIL-BF_4-10	46	4	2	11.5	23.0
PIL-BF_4-20	319	5	3	63.8	106.3
PIL-BF_4-30	708	7	4	101.1	177.0
PIL-TFSI-10	4	2	2	2.0	2.0
PIL-TFSI-20	9	4	3	2.3	3.0
PIL-TFSI-30	254	2	3	127.0	84.7

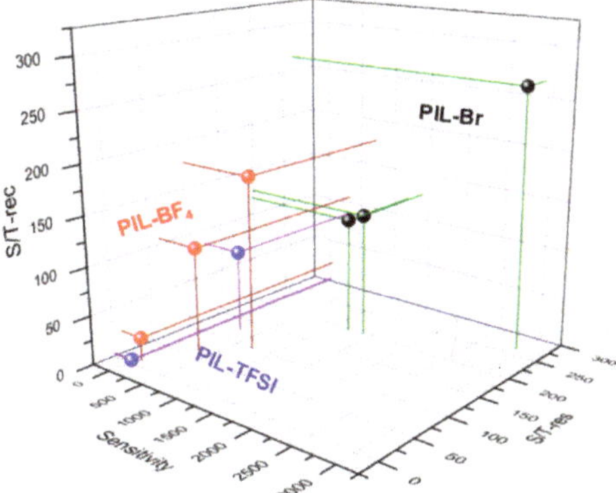

Figure 9. Response of sensors based on PILs(PIL-Br, PIL-BF_4 and PIL-TFSI) at the working frequency.

The PIL structure is special because of the many three-dimensional spaces which are due to the staggered connections between the polymer chains, similar to an overpass-type structure. Meanwhile, these three-dimensional spaces are conducive to increase the free volume by absorbing water molecules for current carrier transport. At this basis, the synergistic effect of three-dimensional space and hydrophilic PIL can significantly improve the humidity performance of sensors. For example, the PIL-based humidity sensors exhibit varying sensitivity from 11% to 98% RH at room temperature (25 °C), as shown in Figure 9. Based on the static water contact angle, the hydrophilicity follows the order PIL-Br > PIL-BF_4 > PIL-TFSI. Among the PILs in our work, the PIL-Br sensor which has the optimum hydrophilicity possesses a biggest free volume based on the swelling theory. Fortunately, the larger free volume can provide more absorption sites to make water molecules permeate into the PIL film for more carrier transport channels at the same ambient humidity and temperature. Due to the electrostatic fields of the PIL-Br film, the absorbed water molecules will dissociate as $2H_2O \rightarrow H_3O^+ + OH^-$. In addition, the conduction in polymer films mainly occurs along the hydrogen bond via hydrogen-ion jumps based on the Grotthuss mechanism, i.e., $H_2O + H_3O^+ \rightarrow H_3O^+ + H_2O$. Thus, the PIL-Br film has more free ions and carrier transport channels to achieve outstanding conductivity, high sensitivity and a large response and recovery velocity, which is verified by our experimental data.

3.5. Possible Mechanism for the Fast Response and Recovery of PILs

As mentioned above, the conductivity of the PIL is inversely proportional to the viscosity and proportional to the free volume through the free volume theory. At the same time, the free volume of PIL has affinity with moisture. Therefore, the mechanism of the fast response and recovery of the PILs can be explained by the free volume theory, as shown in Figure 10. The response process of the PIL-based humidity sensor is described from low RH to high RH as follows: (i) increase of the free volume of PIL film through combining with water molecules to provide more carrier transport channels, (ii) obtain a high concentration of free ions because of the relaxation and disintegrate effect of the polymer chains and dissociation of the absorbed water molecules and (iii) increased conductivity in a PIL film because of the fast free-ion movements in the channels. The PIL structure including anion types and film thickness affected the free volume of the PIL film at the same ambient humidity at 25 °C; then, the hydrophilic anion and slightly thick film benefits an increase in the free volume. In addition, the magnitude of the free volume of the PIL film can decide the lowest humidity required to achieve continuous multi-water layers for providing more carrier transport channels. Meanwhile, the concentration of free ions including H^+, H_3O^+, Br^- and BF_4^- can increase as the ambient humidity increases. Therefore, our findings develop a new perspective to understand the humidity performance of PILs based on free volume theory and we can rationally design PIL structures including anion types and film thickness to obtain super properties including fast response and recovery in PIL sensing films.

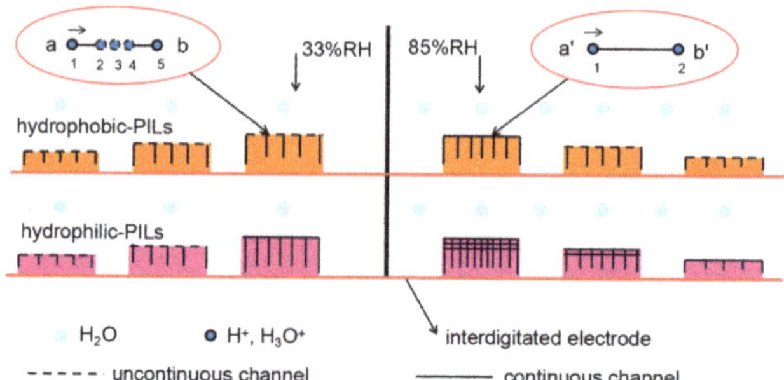

Figure 10. Schematic illustration of the humidity sensing mechanism for a PIL film with varying thickness and anion type.

3.6. Real-Time Monitoring of Human Respiratory Rate

Human breath is a highly complex mixture of more than 100 types of gases, many of which can provide useful information in the monitoring of the human health condition [36]. Water vapour—one of the main ingredients of human breath—is easily detected by a humidity sensor and the change in humidity for respiratory airflow can be used to effectively evaluate human health status. Nowadays, the increase in the incidence of sleep apnea syndrome (SAS) has aroused people's attention to the monitoring of human sleep and has led to a large number of related studies. Apnea during sleep may cause serious consequences and even death. However, this situation will be ameliorated if the sleep apnea of a human is found in a timely manner and appropriate rescue measures are taken.

In recent years, great breakthroughs have been made in research on human respiration monitoring. For example, Yun-Ze Long et al. [37] prepared 1D sensing materials by electrospinning, which were utilized as smart fabrics for avoiding sleep apnea. Youju Huang et al. [38] fabricated a hybrid PNIPAm/AuNP aerogel-based humidity sensor and used it to detect human breath in different states or from different individuals for human health monitoring. Jin Zhou et al. [39] prepared humidity sensors based on a polysquaraine, poly(1-phenylpyrrole-2-ylsquaraine) (PPPS) loaded with varying amounts of Au nanoparticles, which were applied to human respiratory monitoring, showing excellent moisture sensitivity and practicality. In our work, the humidity sensor based on PIL-Br-10 has superior performance, including a fast response, high sensitivity and excellent long stability, and it can be applied in the real-time monitoring of the human respiratory rate. The PIL-Br-10 film could easily detect the change in humidity in respiratory airflow, which could effectively evaluate human health status. Hence, we put the humidity sensor based on the PIL-Br-10 into a mask to fabricate a breath mask, as shown in Figure S9; the sensor was placed at a distance of 3.5 cm from the nose and 2.5 cm from the mouth of a tester. A volunteer (healthy female) wore the breath mask to breathe in a fast, normal or low rate, and the testing results are shown in Figure 11. We can obviously see the distinction for breathing fast, normally or at a low rate, even obtaining a breathing period between 0.9 to 4.3 s, as shown in Figure 11; the frequency of the curve varies with the rate of respiration, revealing that our PIL-Br-10 sensor can capture the signals of humidity change under different rates. This proves that the sensor has practicable application for real-time monitoring of human respiratory rates.

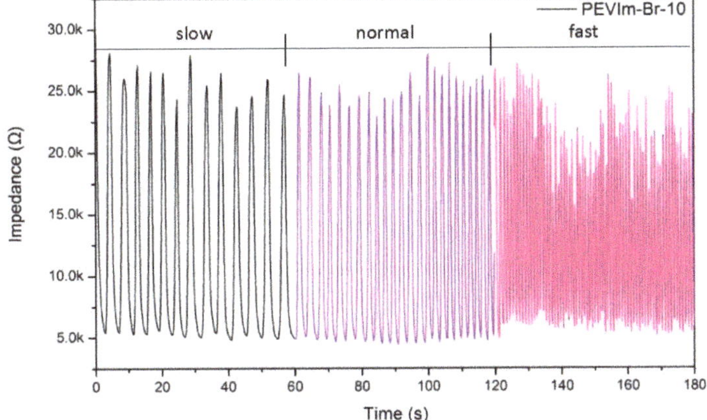

Figure 11. Response of the sensor based on PIL-Br-10 at varying respiratory rates at the working frequency (500 Hz).

4. Conclusions

In summary, PILs were successfully synthesized by typical polymerization and an anion exchange method and were used as humidity materials. We studied the effects of the PIL structure including film thickness and anion type on PIL-based humidity sensor performance. The results showed that a hydrophilic anion and slightly thick film can benefit an increase in the free volume of PIL films to obtain a high concentration of free ions and effective carrier transport channels. Furthermore, the humidity sensing mechanism of PILs was superbly explained by the free volume theory. It is worth noting that our work develops a new perspective to understand the humidity performance of PILs based on free volume theory, resulting in fast response and recovery properties realized by the rational design of PIL sensing films. In our work, the PIL-Br-10 humidity sensor shows high sensitivity (1190), small hysteresis (3.2%), enhanced durability and rapid response (6s)/recovery (10s). We also apply the PIL-Br-10 sensor to monitor human breath under varying conditions, revealing excellent performance and practicability values. Therefore, the PIL-based sensor with novel and reasonable design is highly promising for a broad range of applications.

Supplementary Materials: The following are available online at http://www.mdpi.com/2079-4991/9/5/749/s1.

Author Contributions: Conceptualization, X.D. and J.N.; methodology, S.X.; validation, R.T., J.N. and X.D.; investigation, S.X.; resources, T.W. and X.D.; data curation, R.T. and J.N.; writing—original draft preparation, R.T. and J.N.; writing—review and editing, X.D.; visualization, S.X.; supervision, T.W. and X.D.; project administration, X.D.; funding acquisition, X.D.

Funding: This research was funded by the National Natural Science Foundation of China (grant no. 21601148) and the Natural Science Foundation of Fujian Province (grant no. 2017J05090).

Conflicts of Interest: The authors declare no conflict of interest.

References

1. Tan, Z.Q.; Liu, J.F.; Pang, L. Advances in analytical chemistry using the unique properties of ionic liquids. *TrAC Trends Anal. Chem.* **2012**, *39*, 218–227. [CrossRef]
2. Wishart, J.F. Energy applications of ionic liquids. *Energy Environ. Sci.* **2009**, *2*, 956–961. [CrossRef]
3. Marcilla, R.; Alberto Blazquez, J.; Rodriguez, J.; Pomposo, J.A.; Mecerreyes, D. Tuning the Solubility of Polymerized Ionic Liquids by Simple Anion-Exchange Reactions. *J. Polym. Sci. Part A Polym. Chem.* **2004**, *42*, 208–212. [CrossRef]
4. Kitazawa, Y.; Ueno, K.; Watanabe, M. Advanced Materials Based on Polymers and Ionic Liquids. *Chem. Rec.* **2018**, *18*, 391–409. [CrossRef]

5. Yang, X.; Wang, Y.; Qing, X. A Flexible Capacitive Pressure Sensor Based on Ionic Liquid. *Sensors* **2018**, *18*, 2395. [CrossRef]
6. Kim, S.; Han, S.G.; Koh, Y.G.; Lee, H.; Lee, W. Colorimetric Humidity Sensor Using Inverse Opal Photonic Gel in Hydrophilic Ionic Liquid. *Sensors* **2018**, *18*, 1357. [CrossRef] [PubMed]
7. Li, Y.; Li, G.; Wang, X.; Zhu, Z.; Ma, H.; Zhang, T.; Jin, J. Poly(ionic liquid)-wrapped single-walled carbon nanotubes for sub-ppb detection of CO_2. *Chem. Commun.* **2012**, *48*, 8222–8224. [CrossRef]
8. Yuan, J.; Mecerreyes, D.; Antonietti, M. Poly(ionic liquid)s: An update. *Prog. Polym. Sci.* **2013**, *38*, 1009–1036. [CrossRef]
9. Washiro, S.; Yoshizawa, M.; Nakajima, H.; Ohno, H. Highly ion conductive flexible films composed of network polymers based on polymerizable ionic liquids. *Polymer* **2004**, *45*, 1577–1582. [CrossRef]
10. Vygodskii, Y.S.; Shaplov, A.S.; Lozinskaya, E.I.; Lyssenko, K.A.; Golovanov, D.G.; Malyshkina, I.A.; Gavrilova, N.D.; Buchmeiser, M.R. Conductive Polymer Electrolytes Derived from Poly(norbornene)s with Pendant Ionic Imidazolium Moieties. *Macromol. Chem. Phys.* **2008**, *209*, 40–51. [CrossRef]
11. Zhao, Q.; Yin, M.; Zhang, A.P.; Prescher, S.; Antonietti, M.; Yuan, J. Hierarchically structured nanoporous poly(ionic liquid) membranes: Facile preparation and application in fiber-optic pH sensing. *J. Am. Chem. Soc.* **2013**, *135*, 5549–5552. [CrossRef]
12. Yuan, J.; Antonietti, M. Poly(ionic liquid)s: Polymers expanding classical property profiles. *Polymer* **2011**, *52*, 1469–1482. [CrossRef]
13. Yuan, J.; Antonietti, M. Poly(Ionic Liquid)s as Ionic Liquid-Based Innovative Polyelectrolytes. In *Applications of Ionic Liquids in Polymer Science and Technology*; Springer: Berlin/Heidelberg, Germany, 2015; pp. 47–67.
14. Wang, L.; Duan, X.; Xie, W.; Li, Q.; Wang, T. Highly chemoresistive humidity sensing using poly(ionic liquid)s. *Chem. Commun.* **2016**, *52*, 8417–8419. [CrossRef] [PubMed]
15. Willa, C.; Yuan, J.; Niederberger, M.; Koziej, D. When Nanoparticles Meet Poly(Ionic Liquid)s: Chemoresistive CO_2 Sensing at Room Temperature. *Adv. Funct. Mater.* **2015**, *25*, 2537–2542. [CrossRef]
16. Shang, Y.; Zhang, H.; Wang, X.; Wu, J. An optical olfactory sensor based on porous silicon infiltrated with room-temperature ionic liquid arrays. *Chemistry* **2011**, *17*, 13400–13404. [CrossRef]
17. Kuo, S.M.; Huang, Y.W.; Yeh, S.M.; Cheng, W.H.; Lin, C.H. Liquid crystal modified photonic crystal fiber (LC-PCF) fabricated with an un-cured SU-8 photoresist sealing technique for electrical flux measurement. *Opt. Soc. Am.* **2011**, *19*, 18372–18379. [CrossRef]
18. Ikeda, T.; Moriyama, S.; Kim, J. Imidazolium-based poly(ionic liquid)s with poly(ethylene oxide) main chains: Effects of spacer and tail structures on ionic conductivity. *J. Polym. Sci. Part A Polym. Chem.* **2016**, *54*, 2896–2906. [CrossRef]
19. Hong, S.H.; Tung, T.T.; Kim, T.Y.; Suh, K.S. Preparation of single-walled carbon nanotube (SWNT) gel composites using poly(ionic liquids). *Colloid Polym. Sci.* **2010**, *288*, 1013–1018. [CrossRef]
20. Li, G.; Peng, C.; Zhang, C.; Xu, Z.; Shang, M.; Yang, D.; Kang, X.; Wang, W.; Li, C.; Cheng, Z.; et al. Eu^{3+}/Tb^{3+}-doped $La_2O_2CO_3/La_2O_3$ nano/microcrystals with multiform morphologies: Facile synthesis, growth mechanism, and luminescence properties. *Inorg. Chem.* **2010**, *49*, 10522–10535. [CrossRef]
21. Kim, E.; Kim, S.Y.; Jo, G.; Kim, S.; Park, M.J. Colorimetric and resistive polymer electrolyte thin films for real-time humidity sensors. *ACS Appl. Mater. Interfaces* **2012**, *4*, 5179–5187. [CrossRef] [PubMed]
22. Rubinger, C.P.L.; Martins, C.R.; De Paoli, M.A.; Rubinger, R.M. Sulfonated polystyrene polymer humidity sensor: Synthesis and characterization. *Sens. Actuators B Chem.* **2007**, *123*, 42–49. [CrossRef]
23. Santoso, F.; Redmond, S.J. Indoor location-aware medical systems for smart homecare and telehealth monitoring: State-of-the-art. *Physiol. Meas.* **2015**, *36*, R53. [CrossRef] [PubMed]
24. Laville, C.; Deletage, J.Y.; Pellet, C. Humidity sensors for a pulmonary function diagnostic microsystem. *Sens. Actuators B Chem.* **2001**, *76*, 304–309. [CrossRef]
25. Cortegiani, A.; Sanfilippo, F.; Tramarin, J.; Giarratano, A. Predatory open-access publishing in critical care medicine. *J. Crit. Care* **2019**, *50*, 247–249. [CrossRef]
26. Lee, S.-W.; Choi, B.I.; Kim, J.C.; Woo, S.-B.; Kim, Y.-G.; Kwon, S.; Yoo, J.; Seo, Y.-S. Sorption/desorption hysteresis of thin-film humidity sensors based on graphene oxide and its derivative. *Sens. Actuators B Chem.* **2016**, *237*, 575–580. [CrossRef]
27. Solmaz, R.; Altunbaş Şahin, E.; Döner, A.; Kardaş, G. The investigation of synergistic inhibition effect of rhodanine and iodide ion on the corrosion of copper in sulphuric acid solution. *Corros. Sci.* **2011**, *53*, 3231–3240. [CrossRef]

28. Miyake, T.; Rolandi, M. Grotthuss mechanisms: From proton transport in proton wires to bioprotonic devices. *J. Phys. Condens. Matter* **2016**, *28*, 023001. [CrossRef]
29. Macedo, P.B.; Litovitz, T.A. On the Relative Roles of Free Volume and Activation Energy in the Viscosity of Liquids. *J. Chem. Phys.* **1965**, *42*, 245–256. [CrossRef]
30. Kasapis, C.; Thompson, P.D. The effects of physical activity on serum C-reactive protein and inflammatory markers: A systematic review. *J. Am. Coll. Cardiol.* **2005**, *45*, 1563–1569. [CrossRef]
31. Sopade, P.A. Application of the Williams–Landel–Ferry model to the viscosity–temperature relationship of Australian honeys. *J. Food Eng.* **2003**, *56*, 67–75. [CrossRef]
32. Budzien, J.; McCoy, J.D. Solute mobility and packing fraction: A new look at the Doolittle equation for the polymer glass transition. *J. Chem. Phys.* **2003**, *119*, 9269–9273. [CrossRef]
33. Subbiah, B.; Morison, K.R. Electrical conductivity of viscous liquid foods. *J. Food Eng.* **2018**, *237*, 177–182. [CrossRef]
34. Keddie, J.L.; Jones, R.A.L.; Cory, R.A. Size-Dependent Depression of the Glass-Transition Temperature in Polymer-Films. *Europhys. Lett.* **1994**, *27*, 59–64. [CrossRef]
35. Vila, J.; Ginés, P.; Pico, J.M.; Franjo, C.; Jiménez, E.; Varela, L.M.; Cabeza, O. Temperature dependence of the electrical conductivity in EMIM-based ionic liquids. *Fluid Phase Equilib.* **2006**, *242*, 141–146. [CrossRef]
36. Shin, W.; Goto, T.; Nagai, D.; Itoh, T.; Tsuruta, A.; Akamatsu, T.; Sato, K. Thermoelectric Array Sensors with Selective Combustion Catalysts for Breath Gas Monitoring. *Sensors* **2018**, *18*, 1579. [CrossRef]
37. Zhang, J.; Wang, X.X.; Zhang, B.; Ramakrishna, S.; Yu, M.; Ma, J.W.; Long, Y.Z. In Situ Assembly of Well-Dispersed Ag Nanoparticles throughout Electrospun Alginate Nanofibers for Monitoring Human Breath-Smart Fabrics. *ACS Appl. Mater. Interfaces* **2018**, *10*, 19863–19870. [CrossRef]
38. Ali, I.; Chen, L.; Huang, Y.; Song, L.; Lu, X.; Liu, B.; Zhang, L.; Zhang, J.; Hou, L.; Chen, T. Humidity-Responsive Gold Aerogel for Real-Time Monitoring of Human Breath. *Langmuir* **2018**, *34*, 4908–4913. [CrossRef]
39. Zhou, J.; Xiao, X.; Cheng, X.-F.; Gao, B.-J.; He, J.-H.; Xu, Q.-F.; Li, H.; Li, N.-J.; Chen, D.-Y.; Lu, J.-M. Surface modification of polysquaraines to sense humidity within a second for breath monitoring. *Sens. Actuators B Chem.* **2018**, *271*, 137–146. [CrossRef]

Article

Pressure-Dependent Confinement Effect of Ionic Liquids in Porous Silica

Teng-Hui Wang, En-Yu Lin and Hai-Chou Chang *

Department of Chemistry, National Dong Hwa University, Shoufeng, Hualien 974, Taiwan; 810712101@gms.ndhu.edu.tw (T.-H.W.); 410312036@gms.ndhu.edu.tw (E.-Y.L.)
* Correspondence: hcchang@gms.ndhu.edu.tw; Tel.: +886-3-8903585

Received: 13 March 2019; Accepted: 10 April 2019; Published: 16 April 2019

Abstract: The effect of confining ionic liquids (ILs) such as 1-ethyl-3-methylimidazolium tetrafluoroborate $[C_2C_1Im][BF_4]$ or 1-butyl-3-methylimidazolium tetrafluoroborate $[C_4C_1Im][BF_4]$ in silica matrices was investigated by high-pressure IR spectroscopy. The samples were prepared via the sol-gel method, and the pressure-dependent changes in the C–H absorption bands were investigated. No appreciable changes were observed in the spectral features when the ILs were confined in silica matrices under ambient pressure. That is, the infrared measurements obtained under ambient pressure were not sufficient to detect the interfacial interactions between the ILs and the porous silica. However, dramatic differences were observed in the spectral features of $[C_2C_1Im][BF_4]$ and $[C_4C_1Im][BF_4]$ in silica matrices under the conditions of high pressures. The surfaces of porous silica appeared to weaken the cation-anion interactions caused by pressure-enhanced interfacial IL-silica interactions. This confinement effect under high pressures was less obvious for $[C_4C_1Im][BF_4]$. The size of the cations appeared to play a prominent role in the IL-silica systems.

Keywords: ionic liquid; IR spectroscopy; silica; high pressure

1. Introduction

Porous silica, synthesized using established procedures such as the sol-gel method, have a high surface area and are used in many applications [1–6]. The high surface area property makes porous silica promising candidates for the development of ionogels, biological applications, and the fabrication of mechanical systems [1]. The concept of ionogels is associated with ionic liquids immobilized by solid-like matrices (porous silica, for example) [2–6]. The precursors of porous silica are usually silicon alkoxides such as tetramethyl orthosilicate (TMOS) and tetraethylorthosilicate (TEOS) [5,6].

Ionic liquids (ILs) are salts whose melt temperature is less than 100 °C because of the difficulty of stacking asymmetric and bulky cations and anions [7–17]. Due to the non-volatility and liquid state at room temperature, ILs have been used in energy storage devices, carbon dioxide absorptions, and dye-sensitized solar cells [7–10,12]. Typically, imidazolium-based ionic liquids are most extensively studied, and the cations are characterized by an imidazolium polar head and alkyl tail [7–10]. Aggregation of cations and anions in bulk ILs via Coulombic forces and hydrogen bonding has been proposed, where the cation-anion interactions lead to heterogeneous or organized cluster structures in the IL [7–10].

The confinement effect is a phenomenon wherein molecules or ions are trapped in molecular-dimension rooms (caves in porous materials), where their interactions with the pore-wall surfaces may change the physicochemical behaviors of the molecules or ions [1,8]. The molecules or ions entrapped in porous networks may show physical properties dissimilar to those of the neat liquid states, and the surface interactions with the pores may disturb the molecular or ionic associations [1,8]. As ILs are confined in an inorganic matrix, the mixtures form a two-phase structure, where the solid and liquid phases interconnect throughout the mixture. Much effort has been made to probe the changes in the properties of the ILs, upon physical confinement. These have led to the ILs in immobilized forms, owing to the solid porous matrix [1–6] being applied in a wide variety of applications. Some studies have shown that ILs entrapped in silica matrices (ionogels) have unique properties as compared to the neat ILs [3–6]. The physicochemical nature of the ionogel is determined by the interplay of the cation-anion associations and IL-surface interactions. The confinement of ILs in the host matrix leads to a partial disruption of the cation-anion interactions and cluster structures and cause ion–matrix interactions, resulting in changes to the phase-transition temperatures [3–6]. The confinement effect may also relax the crystallization rate of ILs. In addition to porous silica, some authors proposed that polyvinylidene fluoride (PVDF) and the matrix made from amine hardener and an epoxy prepolymer could act as the host network [11]. Gelled electrolyte-containing ILs may have applications in the future of lithium-ion batteries [12]. Understanding the interfacial structures of ILs at solid surfaces is crucial for extending the application of ILs in energy-storage devices. However, hysteretic anion-cation exchanges on a solid surface (in the first ionic layer) makes it demanding to understand what happens if various surfaces are interfaced with ILs [18,19].

The application of high pressure is an excellent technique to investigate the ordering of ILs on a solid surface [20–22]. The changes in the spectral characteristics induced by high pressures are of particular interest. The local structures of ILs appear disturbed under high pressure, and pressure-enhanced interfacial interactions may occur between the ILs and the solid surfaces, under high pressures. The various degrees of associations between the ILs and the solid surfaces at high pressures may arise from a reorganization of local structures and the hydrogen bonding network [20–23]. Upon compression, the relative weights of the intermolecular forces defining the aggregation states and the intramolecular interactions (molecular bonding) are changed. To obtain a further understanding of the pressure-dependence of the confinement effect [24], we applied high pressure to study the local structures of the ILs confined in porous silica. We note that $P > 30$ GPa is needed to change the electronic structures of the samples. Thus, the pressure used in this study (approximately 2 GPa) mainly affect local structures and interionic distances. Infrared spectroscopy is sensitive to monitor the local structures and potential energy environments, although the crystal and glass conformations may not be conclusively determined by IR. Thus, high-pressure IR spectroscopy may offer a specific means to detect the local structures of ionic liquids in confined geometries.

2. Materials and Methods

1-Ethyl-3-methylimidazolium tetrafluoroborate ($[C_2C_1Im][BF_4]$, 97%, Sigma-Aldrich, St. Louis, MO, USA), 1-butyl-3-methylimidazolium tetrafluoroborate ($[C_4C_1Im][BF_4]$, ≥97.0%, Sigma-Aldrich), tetraethylorthosilicate (TEOS, 99.999%, Sigma-Aldrich), and formic acid (FA, 98–100%, Merck-KGaA, Darmstadt, Germany) were used to prepare the samples. The ionogels (IL/porous silica) were synthesized according to the template method [5,6]. Formic acid and TEOS were mixed in a molar ratio of 8:1, followed by the addition of the IL ($[C_2C_1Im][BF_4]$ or $[C_4C_1Im][BF_4]$) at 5 wt%. The mixture was allowed to gellify for approximately one week. Afterward, the ethanol, ethyl formate, and the remaining FA were removed by vacuum for 24 h. The powder-like samples were dried at 180 °C using a moisture analyzer (MS-70, A&D Company, Tokyo, Japan).

Fourier transformed infrared spectra of the samples were collected on an IR spectrophotometer (Spectrum RXI, Perkin-Elmer, Naperville, IL, USA) equipped with a lithium tantalite (LITA) detector and a 5× beam condenser. A DAC (diamond anvil cell) of the Merrill–Bassett design was used to generate pressures of up to approximately 2 GPa. For high-pressure infrared measurements, two type-IIa diamonds were used. The infrared spectra of empty DAC were measured first in order to remove the infrared absorption of the two diamond anvils. The samples were placed in a 0.3-mm-diameter hole in a metal gasket (thickness 0.25 mm) mounted on the diamond anvil cell. To avoid saturation of the infrared absorption bands, part of the sample-hole was filled with CaF_2 crystals. For each high-pressure spectrum, 1000 scans were collected. The FTIR spectrometer was operated at data point resolution of 2 cm^{-1} (a resolution of 4 cm^{-1}). The pressure calibration was carried out following Wong's method [25,26]. The FTIR spectra of the ionogel samples at ambient pressure were obtained by the CaF_2 pellet method.

3. Results and Discussion

Figure 1 displays the IR spectra of the pure $[C_2C_1Im][BF_4]$ (curve a) and $[C_2C_1Im][BF_4]$ in a silica matrix (curve b) recorded under ambient pressure. The IR spectrum of neat $[C_2C_1Im][BF_4]$ in Figure 1a shows two demarcated peaks at 3124 and 3164 cm^{-1}, corresponding to the C^2–H and $C^{4,5}$–H vibrations, respectively, of the aromatic imidazolium cation [27,28]. There exist three alkyl C–H bands located in the 2925–3025 cm^{-1} region as shown in Figure 1a. The absorption baseline in Figure 1b can be attributed to the Si–OH groups in the silica surface of the ionogel [20]. A comparison of the spectral absorptions in Figures 1b and 1a showed no significant band-shift or feature-change for the C–H vibrational absorptions of the $[C_2C_1Im][BF_4]$ confined in the silica matrix at ambient pressure. The results in Figure 1 indicate that the vibrational spectroscopic measurements of the IL and ionogel, at ambient pressure, were not sufficient to distinguish the IL-silica interactions from the IL-IL associations.

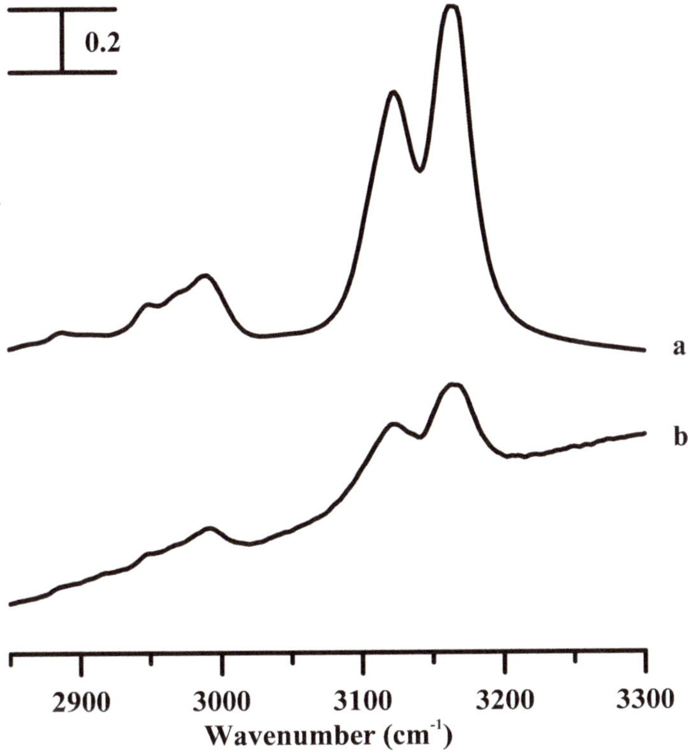

Figure 1. Infrared spectra of the (a) pure $[C_2C_1Im][BF_4]$ and (b) $[C_2C_1Im][BF_4]$ in a silica matrix, recorded at ambient pressure.

Figure 2 shows the IR spectra of neat $[C_2C_1Im][BF_4]$ obtained under ambient pressure (curve a) and at pressures of 0.4 (curve b), 0.7 (curve c), 1.1 (curve d), 1.5 (curve e), 1.8 (curve f), and 2.5 GPa (curve g). In the pressure range from ambient to 0.7 GPa (in Figure 2a–c), the peak broadens in width and blue-shifts in frequency are observed, upon compression, in the C^2–H bands (at ~3124 cm^{-1})

and $C^{4,5}$–H bands (at ~3164 cm^{-1}); these are blue-shifted to 3135 and 3177 cm^{-1}, respectively, at a pressure of 0.7 GPa (Figure 2c). The alkyl C–H bands in the 2925 to 3025 cm^{-1} region also exhibit broadening and blue-shifts with an increase in pressure (≤0.7 GPa), as shown in Figure 2a–c. As the pressure is raised to 1.1 GPa, as in Figure 2d, the wavenumber of the C^2–H stretching band increased to 3143 cm^{-1} with a decrease in the bandwidth. A phase transition or the formation of organized structures may occur, as shown in Figure 2d. As shown in Figure 2d, the $C^{4,5}$–H band (at ~3177 cm^{-1}) is split into three peaks at 3170, 3187, and 3206 cm^{-1} because of the pressure-enhanced interactions. That is, as the $[C_2C_1Im][BF_4]$ is compressed to 1.1 GPa, solid $[C_2C_1Im][BF_4]$ may be present in multiple stable local structures for the $C^{4,5}$–H groups. The imidazolium C–H absorptions in the range between 3100 and 3200 cm^{-1} are complicated by the hydrogen bonding in a cluster mode and Fermi-resonance interactions [27,28]. Previous studies indicated that imidazolium C–H stretching modes of large cluster structures (hydrogen-bonding network) occurred at high wavenumbers [28]. Thus, we assigned the $C^{4,5}$–H bands at 3170, 3187, and 3206 cm^{-1} in Figure 2d to the vibrations of isolated, medium, and large associated local structures, respectively, of the $C^{4,5}$–H groups. The alkyl C–H bands located at ~2967 and 3007 cm^{-1} became dramatically sharp in bandwidth at 1.1 GPa, as shown in Figure 2d. The differences in the spectral absorptions of Figure 2c (0.7 GPa) and d (1.1 GPa) can be related to the generation of an anisotropic environment in Figure 2d, owing to the local alkyl C–H structure changes. It is known that hydrogen bonding leads to ring stacking of ILs and vibrational frequency shifts of imidazolium cations [7–9]. Our results in Figure 2 indicate the important roles played by micro-heterogeneity and hydrogen bonding in $[C_2C_1Im][BF_4]$. As the pressure is increased to 1.5 GPa, the absorption intensity of the $C^{4,5}$–H band at ~3209 cm^{-1} decreases slightly, accompanied by band broadening, as shown in Figure 2e. The decrease in the absorbance of the $C^{4,5}$–H absorption at ~3209 cm^{-1} in Figure 2e may originate from the relaxation of the $C^{4,5}$–H local structures of the pure $[C_2C_1Im][BF_4]$ high-pressure phases upon further compression. The C–H absorptions show continuous band broadening in Figure 2f,g.

Figure 3 shows the IR spectra of $[C_2C_1Im][BF_4]$ in a silica matrix obtained under ambient pressure (curve a) and at pressures of 0.4 (curve b), 0.7 (curve c), 1.1 (curve d), 1.5 (curve e), 1.8 (curve f), and 2.5 GPa (curve g). The $C^{4,5}$–H and C^2–H absorptions show the blue-shifts in frequency to 3174 and 3128 cm^{-1}, respectively, with subtle band broadening in Figure 3c during compression. The aliphatic C–H modes of the alkyl group absorptions display band broadening and a slight blue-shift in the frequency at the pressure of 0.7 GPa in Figure 3c. As the pressure is elevated to 1.1 GPa, the splitting of the $C^{4,5}$–H absorptions is not observed in Figure 3d, unlike the case of the pure $[C_2C_1Im][BF_4]$, being split into three separate bands in Figure 2d. The $C^{4,5}$–H of $[C_2C_1Im][BF_4]$ in a silica matrix shows monotonic blue-shifts in frequency and band-broadening in Figure 3d–g. The alkyl C–H bands in Figure 3d–g also show blue-shifts in frequency and band-broadening, in contrast to the sharp alkyl C-H absorptions in Figure 2d–g. The results in Figure 3d–g suggest that the IL-IL associations have changed as $[C_2C_1Im][BF_4]$ is confined in a silica matrix. The local structures of $[C_2C_1Im][BF_4]$, i.e., the multiple stable local structures, may be perturbed by the silica matrices via the porous surface-IL interactions. In other words, the porous silica may intervene in the organization of the $[C_2C_1Im][BF_4]$ at high pressures (1.1–2.5 GPa), as shown in Figure 3d–g.

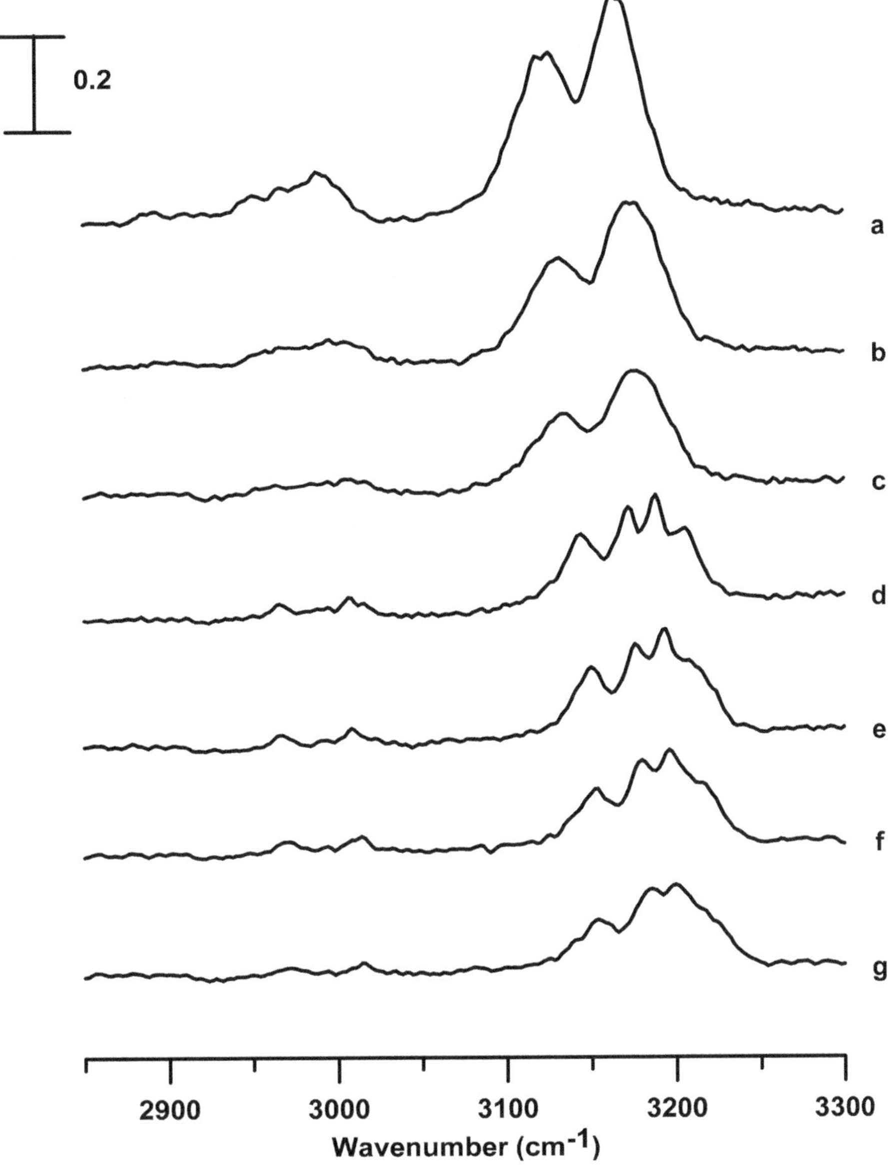

Figure 2. Infrared spectra of the pure $[C_2C_1Im][BF_4]$ obtained at (a) ambient pressure and at (b) 0.4, (c) 0.7, (d) 1.1, (e) 1.5, (f) 1.8, and (g) 2.5 GPa.

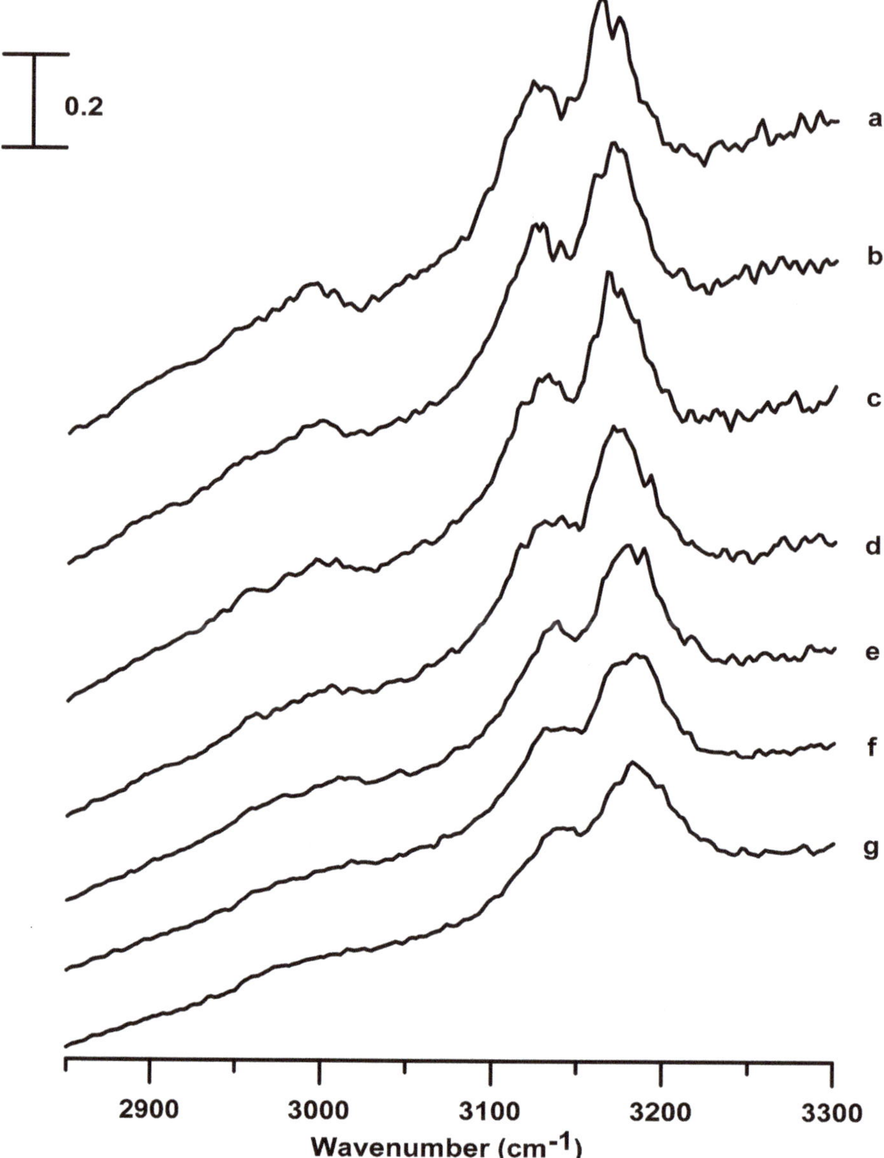

Figure 3. Infrared spectra of $[C_2C_1Im][BF_4]$ in a silica matrix obtained at (a) ambient pressure and at (b) 0.4, (c) 0.7, (d) 1.1, (e) 1.5, (f) 1.8, and (g) 2.5 GPa.

Figure 4 shows the pressure-dependence of the band-shifts in frequency for imidazolium C–H absorptions of the neat $[C_2C_1Im][BF_4]$ and $[C_2C_1Im][BF_4]$ in a silica matrix. The stretching frequencies of the $C^{4,5}$–H (Figure 4a) and C^2–H (Figure 4b) bands at ambient pressure are almost identical for pure $[C_2C_1Im][BF_4]$ and $[C_2C_1Im][BF_4]$ in a silica matrix. At the pressures of 0.4 and 0.7 GPa, the $C^{4,5}$–H (Figure 4a) and C^2–H (Figure 4b) stretching modes undergo mild red-shifts in frequency upon adding the silica matrix. The red-shifts in frequency induced by the presence of the silica matrix became obvious for C^2–H under high pressure (1.1–2.5 GPa), as shown in Figure 4b. Ludwig's group reported that the imidazolium C^2–H stretching band could be described by a major peak at ~3125 cm^{-1} associated with a minor absorption at 3104 cm^{-1}, corresponding to the associated structure and isolated structure, respectively [28]. The absorption components at 3125 cm^{-1} and 3104 cm^{-1} may arise from the large clusters (associated structures) and small clusters (isolated structures), respectively. The decrease in bandwidth of the C^2–H absorption in Figure 2d can be attributed to the decline in the molar ratio of the isolated/associated forms of pure [C_2C_1Im][BF_4] at high pressures. The dramatic decrease in the intensity of the isolated form (shoulder) is not observed in Figure 3d for [C_2C_1Im][BF_4] in the silica matrix under high pressure. Therefore, the red-shifts in frequency in Figure 4b induced under high pressures by the silica matrix for C^2–H may originate from the partial switch of the associated form to the isolated form. These results indicate the prominent role of hydrogen bonding in [C_2C_1Im][BF_4]/silica systems [22,27]. The presence of the silica matrix appears to force the IL to form another arrayed structure, causing the C^2–H isolated structures to increase in intensity under high pressure. We note that the C^2–H associated structures remain the dominant species even at high pressures. As shown in Figure 4a, the $C^{4,5}$–H peak of pure [C_2C_1Im][BF_4] at ~3164 cm^{-1} is blue-shifted upon compression at $P \leq 0.7$ GPa, and is then split into three absorptions (~3206, 3187, and 3170 cm^{-1}) at $P \geq 1.1$ GPa, corresponding to the isolated, medium, and large structures, respectively. Nevertheless, the $C^{4,5}$–H absorptions of [C_2C_1Im][BF_4] in a silica matrix show monotonic blue-shifts in frequency upon compression, as observed in Figure 4a. The top peak positions of $C^{4,5}$–H of [C_2C_1Im][BF_4] in a silica matrix are almost identical to the position of the isolated structures of pure [C_2C_1Im][BF_4]. Consistent with the results of C^2–H, the $C^{4,5}$–H band also reveals an increase in the absorption intensity of the isolated form under high pressures. We note that in contrast to C^2–H, the isolated $C^{4,5}$–H structure is the dominant local conformation for the [C_2C_1Im][BF_4] confined in the silica matrix. According to the experimental results in the literature [5], confinement of [C_2C_1Im][BF_4] may result in the decrease in the dimensionality of solidification of the [C_2C_1Im][BF_4] from a three-dimensional to a one-dimensional structure. Compared to published results [5], our pressure-dependent behaviors observed in Figures 2–4 seem to support the trend of reduced dimensionality for [C_2C_1Im][BF_4] in a silica matrix.

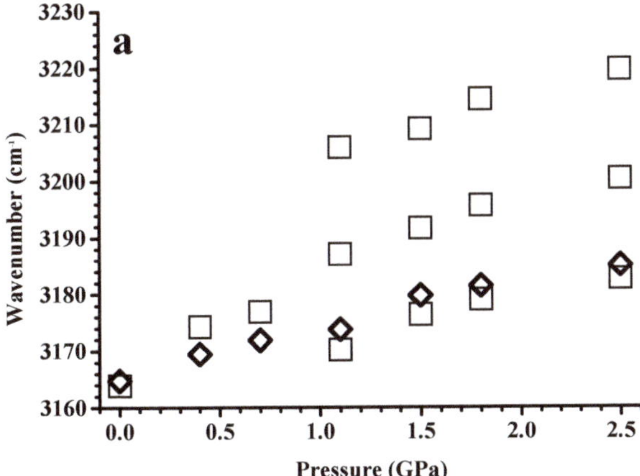

Figure 4. *Cont.*

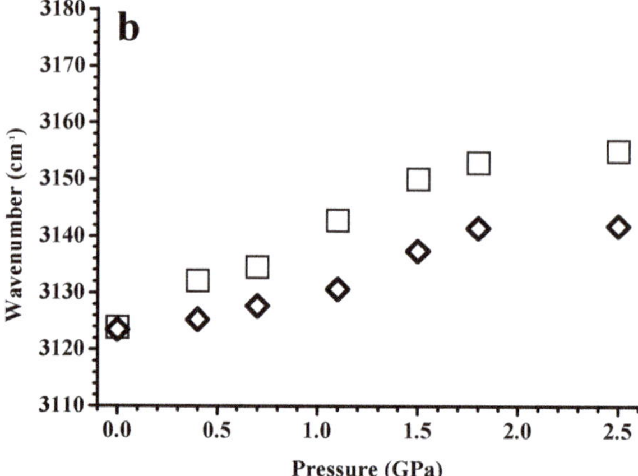

Figure 4. Pressure dependence of the C–H stretching frequencies at (**a**) 3164 and (**b**) 3124 cm^{-1} of the pure $[C_2C_1Im][BF_4]$ (squares) and $[C_2C_1Im][BF_4]$ in a silica matrix (diamonds).

To obtain a clear comprehension into the consequence of the cation size on the pressure-induced isolation/association for pure ILs and IL in a silica matrix, the vibrational spectra of pure $[C_4C_1Im][BF_4]$ and $[C_4C_1Im][BF_4]$ in a silica matrix are studied. Figure S1 (in the Supplementary Materials) shows the IR spectra of the pure $[C_4C_1Im][BF_4]$ (curve a), and $[C_4C_1Im][BF_4]$ in a silica matrix (curve b) recorded under ambient pressure. The IR spectrum of pure $[C_4C_1Im][BF_4]$ in Figure S1a shows two imidazolium C–H peaks at 3120 (C^2–H vibrational bands) and 3163 cm^{-1} ($C^{4,5}$–H vibrational bands), similar to those displayed for the pure $[C_2C_1Im][BF_4]$ in Figure 1a. Three demarcated alkyl C–H bands at 2967, 2939, and 2876 cm^{-1} are observed, caused by the longer alkyl side-chain of the imidazolium of the $[C_4C_1Im][BF_4]$ cation as shown in Figure S1a. Similar to the case in Figure 1, the results obtained at ambient pressure for $[C_4C_1Im][BF_4]$ and $[C_4C_1Im][BF_4]$ in a silica matrix in Figure S1 are not sensitive enough to study the isolated/associated forms of pure $[C_4C_1Im][BF_4]$ and $[C_4C_1Im][BF_4]$ in a silica matrix.

Figure 5 shows the IR spectra of the pure $[C_4C_1Im][BF_4]$ obtained under ambient pressure (curve a) and at pressures of 0.4 (curve b), 0.7 (curve c), 1.1 (curve d), 1.5 (curve e), 1.8 (curve f), and 2.5 GPa (curve g). As the pressure is increased (ambient–2.5 GPa), the C^2–H (3120 cm^{-1}), $C^{4,5}$–H (3163 cm^{-1}), and alkyl C–H (2967 cm^{-1}) absorptions are blue-shifted in frequency accompanied by band broadening, in the case of the pure $[C_4C_1Im][BF_4]$, as shown in Figure 5. Upon compression to high pressures (ambient–2.5 GPa) in Figure 5, the $C^{4,5}$–H band does not show the band-splitting, and the C^2–H band does not reveal band-narrowing, as observed in Figure 2 for the pure $[C_2C_1Im][BF_4]$. That is, the pure $[C_4C_1Im][BF_4]$ may form a less organized structure at high pressures. This phenomenon of forming a less organized structure, presented in Figure 5, is quite different from that of forming organized structures (in the case of pure $[C_2C_1Im][BF_4]$ in Figure 2), which can be attributed to the longer alkyl side-chains of the aromatic ring of the $[C_4C_1Im][BF_4]$ cation. That is, the possibility of the organization is more for symmetric cations such as $[C_2C_1Im]$ than for asymmetrical structures, such as $[C_4C_1Im]$.

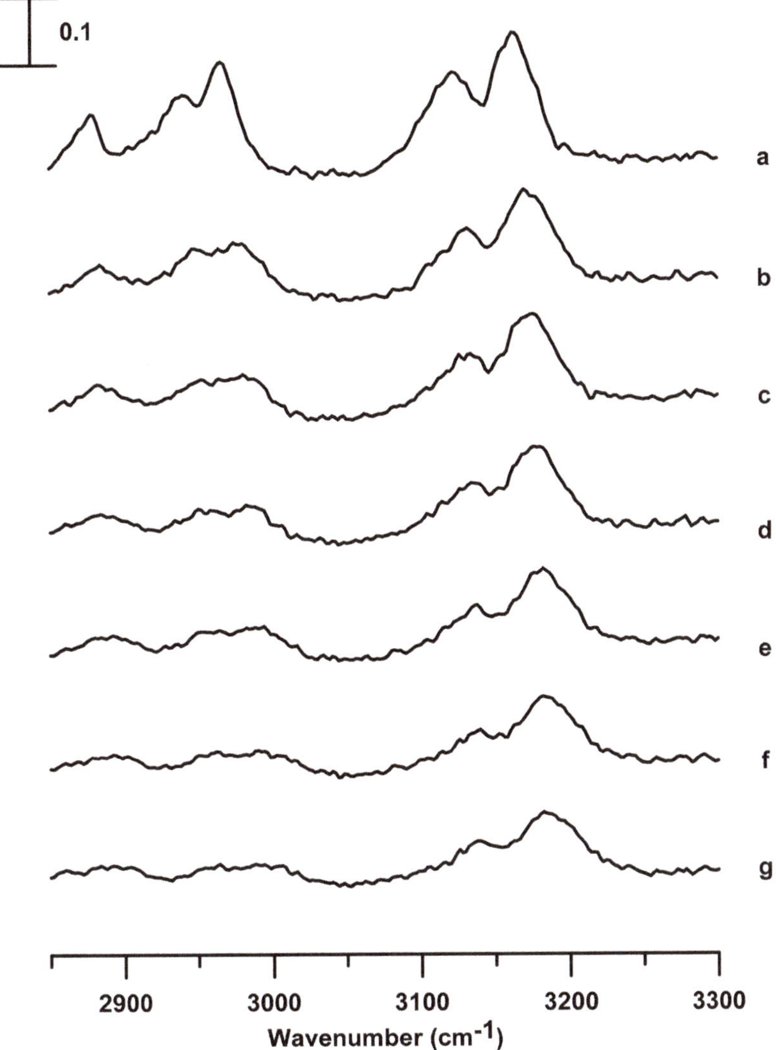

Figure 5. Infrared spectra of the pure $[C_4C_1Im][BF_4]$ obtained at (a) ambient pressure and at (b) 0.4, (c) 0.7, (d) 1.1, (e) 1.5, (f) 1.8, and (g) 2.5 GPa.

Figure 6 shows the IR absorptions of $[C_4C_1Im][BF_4]$ in a silica matrix obtained under ambient pressure (curve a) and at pressures of 0.4 (curve b), 0.7 (curve c), 1.1 (curve d), 1.5 (curve e), 1.8 (curve f), and 2.5 GPa (curve g). The spectral features of $[C_4C_1Im][BF_4]$ in a silica matrix, shown in Figure 6, are similar to those displayed for pure $[C_4C_1Im][BF_4]$ in Figure 5. In Figure 6, which presents the results of $[C_4C_1Im][BF_4]$ in a silica matrix, the $C^{4,5}$–H (3164 cm^{-1}), C^2–H (3120 cm^{-1}), and alkyl C–H (2967 cm^{-1}) bands are blue-shifted accompanied by band broadening, as the pressure is surged from ambient to 2.5 GPa.

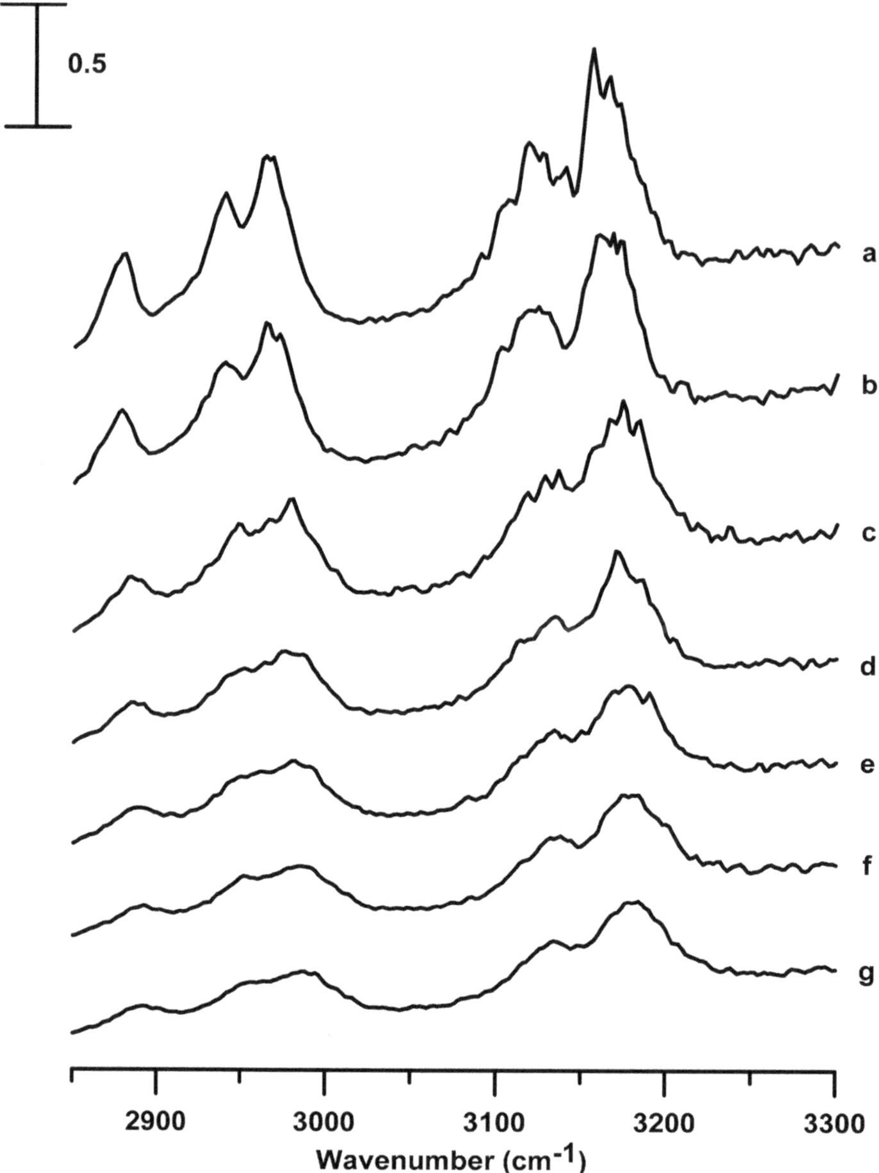

Figure 6. Infrared spectra of $[C_4C_1Im][BF_4]$ in a silica matrix obtained at (a) ambient pressure and at (b) 0.4, (c) 0.7, (d) 1.1, (e) 1.5, (f) 1.8, and (g) 2.5 GPa.

Figure 7 shows the pressure-dependence of the band-shifts in frequencies for imidazolium C–H absorptions of pure $[C_4C_1Im][BF_4]$ and $[C_4C_1Im][BF_4]$ in a silica matrix. Figure S2 in the Supplementary Materials shows the pressure-dependence of the band-shifts of the alkyl C–H absorptions at ~2967 cm^{-1} for the pure $[C_4C_1Im][BF_4]$ and $[C_4C_1Im][BF_4]$ in a silica matrix. As shown in Figure 7 and Figure S2, remarkable blue-shifts in frequency occur upon compression to 0.4 GPa for pure $[C_4C_1Im][BF_4]$. This observation suggests a phase transition of pure $[C_4C_1Im][BF_4]$ at 0.4 GPa. For $[C_4C_1Im][BF_4]$

in a silica matrix, initially, the C–H bands shift slightly in frequency (or show almost no change) at 0.4 GPa, but blue-shifts in frequency occur upon further increase of the pressure to 0.7 GPa, as shown in Figure 7 and Figure S2. These results indicate that the presence of the silica matrix may inhibit the solidification of $[C_4C_1Im][BF_4]$ via pressure-enhanced interfacial interactions. The C–H bands show mild frequency-shifts upon further compression for both pure $[C_4C_1Im][BF_4]$ and $[C_4C_1Im][BF_4]$ in a silica matrix. Minor differences are observed in the C–H vibrational frequency, between pure $[C_4C_1Im][BF_4]$ and $[C_4C_1Im][BF_4]$ in a silica matrix, under high pressures, as observed in Figure 7 and Figure S2. In other words, silica can be employed to change the hydrogen-bonded structure of $[C_4C_1Im][BF_4]$, as the pressure is increased. However, the changes in the aggregation behavior of $[C_4C_1Im][BF_4]$ via pressure-enhanced IL-surface interactions are less obvious than those of $[C_2C_1Im][BF_4]$.

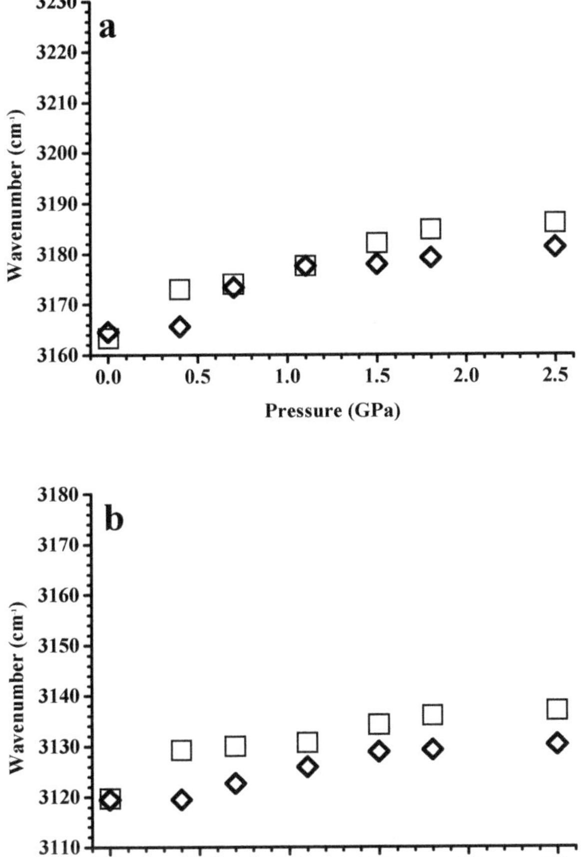

Figure 7. Pressure dependence of the C-H stretching frequencies at (**a**) 3163 and (**b**) 3120 cm^{-1} of the pure $[C_4C_1Im][BF_4]$ (squares) and $[C_4C_1Im][BF_4]$ in a silica matrix (diamonds).

To develop insight into the effect of anions, the preliminary pressure-dependent results of pure $[C_4C_1Im][PF_6]$ and $[C_4C_1Im][PF_6]$ in a silica matrix are displayed in Figures S3 and S4, respectively, in the Supplementary Materials. Figure S5 (in Supplementary Materials) shows pressure dependence of the C–H stretching frequencies at 3170 cm^{-1} of the pure $[C_4C_1Im][PF_6]$ and $[C_4C_1Im][PF_6]$ in a silica matrix, and the changes in aggregation behavior of $[C_4C_1Im][PF_6]$ caused by silica are revealed in Figure S5.

4. Conclusions

We conclude that ILs with short alkyl side-chains ($[C_2C_1Im][BF_4]$) tended to form organized structures and that local C–H structures were affected significantly by the surface-IL interactions at high pressures. Both isolated and associated conformations existed in the pure $[C_2C_1Im][BF_4]$. However, the associated form partially dissociated into isolated structures caused by pressure-enhanced IL-silica interfacial interactions. On the other hand, compared to $[C_2C_1Im][BF_4]$, the asymmetric $[C_4C_1Im][BF_4]$ could form a less organized solid, and the confinement effect induced by the silica matrix, upon compression, became less obvious for $[C_4C_1Im][BF_4]$ in a silica matrix. We note that evidence on the crystalline or glass-like state of ionic liquids can be offered by diffraction. Therefore, a sensitive detection technique such as diffraction may be helpful for future applications.

Supplementary Materials: The following are available online at http://www.mdpi.com/2079-4991/9/4/620/s1, Figure S1: Infrared spectra of (a) pure $[C_4C_1Im][BF_4]$ and (b) $[C_4C_1Im][BF_4]$ in a silica matrix, recorded at ambient pressure, Figure S2: Pressure dependence of the C–H stretching frequencies at 2967 cm^{-1} of the pure $[C_4C_1Im][BF_4]$ (squares) and $[C_4C_1Im][BF_4]$ in a silica matrix (diamonds). Figure S3: Infrared spectra of the pure $[C_4C_1Im][PF_6]$ obtained at (a) ambient pressure and at (b) 0.4, (c) 0.7, (d) 1.1, (e) 1.5, (f) 1.8, (g) 2.5 GPa, and (h) back to ambient. Figure S4: Infrared spectra of $[C_4C_1Im][PF_6]$ in a silica matrix obtained at (a) ambient pressure and at (b) 0.4, (c) 0.7, (d) 1.1, (e) 1.5, (f) 1.8, and (g) 2.5 GPa. Figure S5: Pressure dependence of the C–H stretching frequencies at 3170 cm^{-1} of the pure $[C_4C_1Im][PF_6]$ (squares) and $[C_4C_1Im][PF_6]$ in a silica matrix (diamonds).

Author Contributions: Formal analysis and writing of the original draft—T.-H.W.; Data curation—E.-Y.L.; Supervision, writing-review, and editing—H.-C.C.

Funding: This research was funded by the Ministry of Science and Technology (Contract No. MOST 107-2113-M-259-005) of Taiwan.

Acknowledgments: The authors thank the National Dong Hwa University and Ministry of Science and Technology for financial support.

Conflicts of Interest: The authors declare no conflict of interest.

References

1. Zhang, W.; Zhao, Q.; Yuan, J. Porous Polyelectrolytes: The Interplay of Charge and Pores for New Functionalities. *Angew. Chem.* **2018**, *57*, 6754–6773. [CrossRef] [PubMed]
2. Yasui, T.; Kamio, E.; Matsuyama, H. Inorganic/Organic Double-Network Ion Gels with Partially Developed Silica-Particle Network. *Langmuir* **2018**, *34*, 10622–10633. [CrossRef]
3. Garaga, M.N.; Dracopoulos, V.; Werner-Zwanziger, U.; Zwanziger, J.W.; Maréchal, M.; Persson, M.; Nordstierna, L.; Martinelli, A. A Long-Chain Protic Ionic Liquid inside Silica Nanopores: Enhanced Proton Mobility Due to Efficient Self-Assembly and Decoupled Proton Transport. *Nanoscale* **2018**, *10*, 12337–12348. [CrossRef]
4. Canova, F.F.; Mizukami, M.; Imamura, T.; Kurihara, K.; Shluger, A.L. Structural Stability and Polarisation of Ionic Liquid Films on Silica Surfaces. *Phys. Chem. Chem. Phys.* **2015**, *17*, 17661–17669. [CrossRef]
5. Gupta, A.K.; Singh, R.K.; Chandra, S. Crystallization Kinetics Behavior of Ionic Liquid [Emim] $[BF_4]$ Confined in Mesoporous Silica Matrices. *RSC Adv.* **2014**, *4*, 22277–22287. [CrossRef]
6. Gupta, A.K.; Singh, M.P.; Singh, R.K.; Chandra, S. Low Density Ionogels Obtained by Rapid Gellification of Tetraethyl Orthosilane Assisted by Ionic Liquids. *Dalton Trans.* **2012**, *41*, 6263–6271. [CrossRef]
7. Hayes, R.; Warr, G.G.; Atkin, R. Structure and Nanostructure in Ionic Liquids. *Chem. Rev.* **2015**, *115*, 6357–6426. [CrossRef] [PubMed]
8. Zhang, S.; Zhang, J.; Zhang, Y.; Deng, Y. Nanoconfined Ionic Liquids. *Chem. Rev.* **2016**, *117*, 6755–6833. [CrossRef]
9. Kiefer, J. Optics and Spectroscopy for Fluid Characterization. *Appl. Sci.* **2018**, *8*, 828. [CrossRef]

10. He, Z.; Alexandridis, P. Nanoparticles in Ionic Liquids: Interactions and Organization. *Phys. Chem. Chem. Phys.* **2015**, *17*, 18238–18261. [CrossRef] [PubMed]
11. Lins, L.C.; Livi, S.; Maréchal, M.; Duchet-Rumeau, J.; Gérard, J.-F. Structural Dependence of Cations and Anions to Building the Polar Phase of PVdF. *Eur. Polym. J.* **2018**, *107*, 236–248. [CrossRef]
12. Leclère, M.; Bernard, L.; Livi, S.; Bardet, M.; Guillermo, A.; Picard, L.; Duchet-Rumeau, J. Gelled Electrolyte Containing Phosphonium Ionic Liquids for Lithium-Ion Batteries. *Nanomaterials* **2018**, *8*, 435. [CrossRef] [PubMed]
13. Liu, F.; Zhong, X.; Xu, J.; Kamali, A.; Shi, Z. Temperature Dependence on Density, Viscosity, and Electrical Conductivity of Ionic Liquid 1-Ethyl-3-Methylimidazolium Fluoride. *Appl. Sci.* **2018**, *8*, 356. [CrossRef]
14. Kausteklis, J.; Aleksa, V.; Iramain, M.A.; Brandán, S.A. Dft and Vibrational Spectroscopy Study of 1-Butyl-3-Methylimidazolium Trifluoromethanesulfonate Ionic Liquid. *J. Mol. Struct.* **2019**, *1175*, 663–676. [CrossRef]
15. Shah, F.U.; Holmgren, A.; Rutland, M.W.; Glavatskih, S.; Antzutkin, O.N. Interfacial Behavior of Orthoborate Ionic Liquids at Inorganic Oxide Surfaces Probed by Nmr, Ir, and Raman Spectroscopy. *J. Phys. Chem.* **2018**, *122*, 19687–19698. [CrossRef]
16. Vicent-Luna, J.M.; Dubbeldam, D.; Gómez-Álvarez, P.; Calero, S. Aqueous Solutions of Ionic Liquids: Microscopic Assembly. *ChemPhysChem.* **2016**, *17*, 380–386. [CrossRef]
17. Babucci, M.; Balci, V.; Akçay, A.; Uzun, A. Interactions of [Bmim][BF_4] with Metal Oxides and Their Consequences on Stability Limits. *J. Phys. Chem.* **2016**, *120*, 20089–20102. [CrossRef]
18. Motobayashi, K.; Osawa, M. Recent Advances in Spectroscopic Investigations on Ionic Liquid/Electrode Interfaces. *Curr. Opin. Electrochem.* **2018**, *8*, 147–155. [CrossRef]
19. Motobayashi, K.; Nishi, N.; Inoue, Y.; Minami, K.; Sakka, T.; Osawa, M. Potential-Induced Restructuring Dynamics of Ionic Liquids on a Gold Electrode: Steric Effect of Constituent Ions Studied by Surface-Enhanced Infrared Absorption Spectroscopy. *J. Electroanal. Chem.* **2017**, *800*, 126–133. [CrossRef]
20. Chang, H.-C.; Wang, T.-H.; Burba, C. Probing Structures of Interfacial 1-Butyl-3-Methylimidazolium Trifluoromethanesulfonate Ionic Liquid on Nano-Aluminum Oxide Surfaces Using High-Pressure Infrared Spectroscopy. *Appl. Sci.* **2017**, *7*, 855. [CrossRef]
21. Chang, H.-C.; Hsu, D.-T. Interactions of Ionic Liquids and Surfaces of Graphene Related Nanoparticles under High Pressures. *Phys. Chem. Chem. Phys.* **2017**, *19*, 12269–12275. [CrossRef] [PubMed]
22. Chang, Y.-H.; Chang, H.-C.; Fu, Y.-P. Utilizing Infrared Spectroscopy to Analyze the Interfacial Structures of Ionic Liquids/Al_2O_3 and Ionic Liquids/Mica Mixtures under High Pressures. *Nanomaterials* **2019**, *9*, 373. [CrossRef] [PubMed]
23. Chang, H.-C.; Zhang, R.-L.; Hsu, D.-T. The Effect of Pressure on Cation–Cellulose Interactions in Cellulose/Ionic Liquid Mixtures. *Phys. Chem. Chem. Phys.* **2015**, *17*, 27573–27578. [CrossRef]
24. Adrjanowicz, K.; Szklarz, G.; Koperwas, K.; Paluch, M. Comparison of High Pressure and Nanoscale Confinement Effects on Crystallization of the Molecular Glass-Forming Liquid, Dimethyl Phthalate. *Phys. Chem. Chem. Phys.* **2017**, *19*, 14366–14375. [CrossRef] [PubMed]
25. Wong, P.; Moffatt, D. The Uncoupled OH or OD Stretch in Water as an Internal Pressure Gauge for High-Pressure Infrared Spectroscopy of Aqueous Systems. *Appl. Spectrosc.* **1987**, *41*, 1070–1072. [CrossRef]
26. Wong, P.; Moffatt, D.; Baudais, F. Crystalline Quartz as an Internal Pressure Calibrant for High-Pressure Infrared Spectroscopy. *Appl. Spectrosc.* **1985**, *39*, 733–735. [CrossRef]
27. Roth, C.; Chatzipapadopoulos, S.; Kerlé, D.; Friedriszik, F.; Lütgens, M.; Lochbrunner, S.; Kühn, O.; Ludwig, R. Hydrogen Bonding in Ionic Liquids Probed by Linear and Nonlinear Vibrational Spectroscopy. *New J. Phys.* **2012**, *14*, 105026. [CrossRef]
28. Köddermann, T.; Wertz, C.; Heintz, A.; Ludwig, R. Ion-Pair Formation in the Ionic Liquid 1-Ethyl-3-Methylimidazolium Bis (Triflyl) Imide as a Function of Temperature and Concentration. *ChemPhysChem.* **2006**, *7*, 1944–1949. [CrossRef]

Article

A Cross-Linker-Based Poly(Ionic Liquid) for Sensitive Electrochemical Detection of 4-Nonylphenol

Jian Hu [1,2], Hao Dai [1,2], Yanbo Zeng [2,*], Yiwen Yang [2], Hailong Wang [2], Xudong Zhu [2], Lei Li [2,*], Guobao Zhou [2], Ruoyu Chen [1,*] and Longhua Guo [3]

1 School of Petrochemical Engineering, Changzhou University, Changzhou 213016, China; 15189761689@163.com (J.H.); 13257926215@163.com (H.D.)
2 College of Biological, Chemical Sciences and Engineering, Jiaxing University, Jiaxing 314001, China; yangyiwen@mail.zjxu.edu.cn (Y.Y.); wanghailong@mail.zjxu.edu.cn (H.W.); zhuxudong9zxd@163.com (X.Z.); gbzhou@mail.zjxu.edu.cn (G.Z.)
3 MOE Key Laboratory for Analytical Science of Food Safety and Biology, Fujian Provincial Key Laboratory of Analysis and Detection Technology for Food Safety, Institute of Nanomedicine and Nanobiosensing, College of Chemistry, Fuzhou University, Fuzhou 350116, China; guolh@fzu.edu.cn
* Correspondence: ybzeng@mail.zjxu.edu.cn (Y.Z.); lei.li@mail.zjxu.edu.cn (L.L.); cxdcry@163.com (R.C.); Tel.: +86-573-83646203 (Y.Z.)

Received: 15 February 2019; Accepted: 28 March 2019; Published: 2 April 2019

Abstract: In this study, we report a cross-linker-based poly(ionic liquid) (PIL) for the sensitive detection of 4-nonylphenol (4-NP). PIL was poly(1,4-butanediyl-3,3′-bis-l-vinylimidazolium dibromide) (poly([$V_2C_4(mim)_2]Br_2$)). Poly([$V_2C_4(mim)_2]Br_2$) was prepared via one-step free-radical polymerization. The poly([$V_2C_4(mim)_2]Br_2$) was characterized by infrared spectroscopy, Raman spectroscopy, thermal gravimetric analyzer and scanning electron microscope. The poly([$V_2C_4(mim)_2]Br_2$) was then drop-cast onto a glassy carbon electrode (GCE) to obtain poly([$V_2C_4(mim)_2]Br_2$)/GCE. In comparison with a bare GCE, poly([$V_2C_4(mim)_2]Br_2$)/GCE exhibited higher peak current responses for $[Fe(CN)_6]^{3-/4-}$, lower charge transfer resistance, and larger effective surface area. While comparing the peak current responses, we found the peak current response for 4-NP using poly([$V_2C_4(mim)_2]Br_2$)/GCE to be 3.6 times higher than a traditional cross-linker ethylene glycol dimethacrylate (EGDMA) based poly(EGDMA) modified GCE. The peak current of poly([$V_2C_4(mim)_2]Br_2$) sensor was linear to 4-NP concentration from 0.05 to 5 μM. The detection limit of 4-NP was obtained as 0.01 μM (S/N = 3). The new PIL based electrochemical sensor also exhibited excellent selectivity, stability, and reusability. Furthermore, the poly([$V_2C_4(mim)_2]Br_2$)/GCE demonstrated good 4-NP detection in environmental water samples.

Keywords: poly(ionic liquid); cross-linker; electrochemical detection; 4-nonylphenol

1. Introduction

Nonylphenol ethoxylates are widely used for manufacturing various commercial products such as detergents, lubricating oil, paints, pesticides, and emulsifier, etc [1]. Nonylphenol ethoxylates degrade to produce 4-nonylphenol (4-NP) in an aquatic environment [2]. Since 4-NP can be produced industrially, naturally and through biodegradation of alkylphenol ethoxylates, the high prevalence of 4-NP in environment has been a grave concern due to its ability to mimic estrogen activity [2]. It is a well-known endocrine disruptor and xenoestrogen, causing serious harm to the reproductive health of human and wildlife [3]. Therefore, detection of 4-NP is of great significance. Various analytical methods including enzyme-linked immunosorbent assay [4], liquid chromatography [5], gas chromatography [6], fluorescence analysis [7] and electrochemical method [8] have been developed for the determination of 4-NP. Among these methods, the electrochemical method has received an increasing attention due to its many advantages such as time-saving, low cost, high sensitivity, and real-time detection.

Poly(ionic liquids)s (PILs), a sub-class of polyelectrolytes, are polymeric materials in which the polymer backbone contains an ionic liquid (IL) species in each repeating unit [9]. Therefore, PILs are smart materials having the advantages of ILs (enhanced ionic conductivity and high electrochemical stability) combined with excellent processability and good mechanical performances and resulting from the polymeric structure [10,11]. Owing to such unique combination of properties, PILs find applications [12] in diverse fields such as electrochemical supercapacitors [13], catalyst [14], gas adsorption [15], extraction [16–19], capillary electrochromatography [20], electrochemical sensors [21,22] and fluorescent determination [23]. In recent years, PIL-based materials prepared with IL monomer as modifiers for electrochemical sensors have been found to exhibit high sensitivity for analytes [24–27]. For example, Wang and co-workers prepared graphene oxide-poly(1-[3-(N-pyrrolyl) propyl]-3-butylimidazolium bromide) and used it as a modifier to construct an electrochemical sensor of bisphenol A detection [24]. In our previous work, we reported an electrochemical sensor with the composite of reduced graphene oxide and poly(1-vinyl-3-ethylimidazolium tetrafluoroborate) for sensitive detection of phenylethanolamine A [26]. Yu et al. utilized the composite of dual hydroxyl-functionalized poly (ionic liquid) and 2,2′-Azinobis-(3-ethylbenzthiazoline-6-sulfonate) as the support to construct a ratiometric electrochemical biosensor for Cu^{2+} determination [25].

Although monomer-based PIL composites as modifiers for electrochemical sensors have been extensively investigated [24–27], cross-linker based PILs or only PILs without any functionalization as modifiers for electrochemical sensors have rarely been studied [28–31]. This prompted us to carry out the current research, where we prepared a cross-linker-based PIL poly(1,4-butanediyl-3,3′-bis-l-vinylimidazolium dibromide) (poly($[V_2C_4(mim)_2]Br_2$)) for sensitive electrochemical detection of 4-NP. Poly($[V_2C_4(mim)_2]Br_2$) was synthesized using a simple one-step free-radical polymerization method. The peak current responses of 4-NP using poly($[V_2C_4(mim)_2]Br_2$) and a traditional cross-linker ethylene glycol dimethacrylate based poly (EGDMA) modified glassy carbon electrodes were investigated. Apart from selectivity, stability and reproducibility, we applied poly($[V_2C_4(mim)_2]Br_2$) sensor for 4-NP detection in environmental samples.

2. Materials and Methods

2.1. Materials

4-Nonylphenol and 2,2′-azobisisobutyronitrile (AIBN) were obtained from Aladdin Industrial Corporation (Shanghai, China). Ethylene glycol dimethacrylate (EGDMA), 1-vinylimidazole, 1,4-dibromobutane, 2-nitroaniline, 3-nitroaniline, and 4-nitroaniline were procured from Sigma-Aldrich (St. Louis, MO, USA). IL $[V_2C_4(mim)_2]Br_2$ was prepared in our laboratory. Chitosan was procured from Sinopharm Chemical Regent Co., Ltd. (Shanghai, China). Phosphate buffer was prepared using NaH_2PO_4 and Na_2HPO_4. Deionized water of 18 MΩ cm was applied in the experiments.

2.2. Instrumentation

Fourier transform infrared (FT-IR) spectra were obtained with Nexus-470 FT-IR spectrometer (Nicolet, Madison, USA). Surface morphology measurements were performed using a JSM-7500F scanning electron microscope (SEM) (JEOL, Tokyo, Japan). Raman spectra were acquired on a Thermo Scientific DXR2xi Raman spectrometer (Waltham, MA, USA). Thermal gravimetric analysis (TGA) was obtained on a STA-449F3 instrument (Netzsch, Selb, Germany). All electrochemical experiments were accomplished on a CHI660D electrochemical workstation using a three-electrode system, where a modified glassy carbon electrode (GCE) as the working electrode, a platinum wire electrode used as the auxiliary electrode and a saturated calomel electrode (SCE) as the reference electrode.

2.3. Preparation of Poly(Ionic Liquid) with $[V_2C_4(mim)_2]Br_2$ as Cross-Linker

Scheme 1 shows the synthesis route of poly($[V_2C_4(mim)_2]Br_2$) as a PIL, involving two steps. IL $[V_2C_4(mim)_2]Br_2$ was first synthesized from 1-vinylimidazole and 1,4-dibromo butane following a previously published method [30,31]. Poly($[V_2C_4(mim)_2]Br_2$) was then prepared using $[V_2C_4(mim)_2]Br_2$ as a cross-linker. $[V_2C_4(mim)_2]Br_2$ (0.01 mol) and 50 mg of AIBN were added to 30 mL of mixed solvent containing acetonitrile and toluene (30 mL, $V_{acetonitrile}:V_{toluene} = 1:1$). The reaction mixture was purged with nitrogen for 30 min and allowed to polymerize for 24 h at 60 °C. The obtained product was washed with acetonitrile three times and dried at 60 °C under vacuum overnight to obtain a white solid of poly($[V_2C_4(mim)_2]Br_2$). For comparison, poly(EGDMA) with traditional cross-linker was synthesized following the same procedure, only replacing $[V_2C_4(mim)_2]Br_2$ by EGDMA.

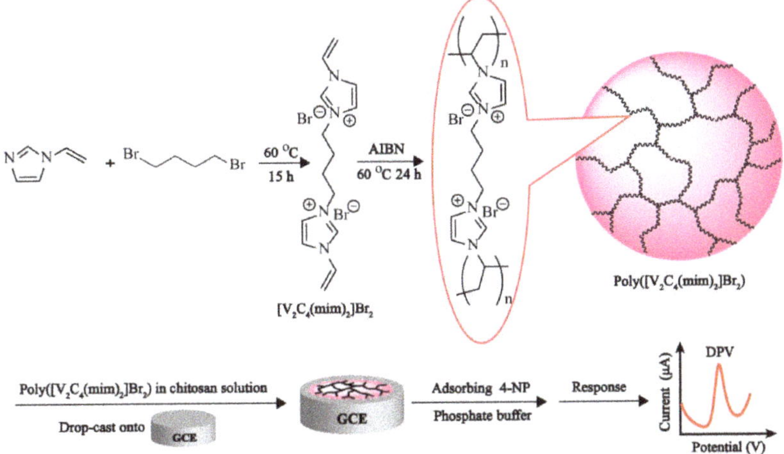

Scheme 1. The synthesis route of poly($[V_2C_4(mim)_2]Br_2$) and the 4-NP detection process using poly($[V_2C_4(mim)_2]Br_2$)/GCE.

2.4. Electrochemical Measurements

Scheme 1 shows the 4-NP detection process using poly($[V_2C_4(mim)_2]Br_2$) as an electrochemical modifier. Poly($[V_2C_4(mim)_2]Br_2$) (3.0 mg) was dispersed in HAc (1 mL, 1 M) containing chitosan (0.5 wt%) by ultrasonication for 30 min. Each suspension (3.0 µL) was drop-cast onto a cleaned GCE and dried under an infrared lamp. Chitosan was applied as an adhesive to stabilize poly($[V_2C_4(mim)_2]Br_2$) onto GCE. The modified GCE were incubated with 4-NP in 0.1 M phosphate buffer (10 mL, pH 6.0) for 6 min and measured by DPV from 0.2 to 0.9 V.

3. Results

3.1. Characterization of Poly($[V_2C_4(mim)_2]Br_2$)

Figure 1A illustrates FT-IR spectra of $[V_2C_4(mim)_2]Br_2$, poly($[V_2C_4(mim)_2]Br_2$), EGDMA and poly(EGDMA). The peak at 1634 cm^{-1} in the FT-IR spectrum of $[V_2C_4(mim)_2]Br_2$ corresponds to exocyclic C=C stretching vibration. The characteristic peaks at 1500, 1401, and 1160 cm^{-1} for $[V_2C_4(mim)_2]Br_2$ are attributed to C=N, C=C, and C-N bonds of the imidazolium cation vibrations [32]. The peaks at 1723, 1636, 1295 and 1153 cm^{-1} are ascribed to C-O stretching vibration of carboxylic ester, C=C, C–O stretching vibration of symmetric and asymmetric ester [33], respectively. Poly($[V_2C_4(mim)_2]Br_2$) and poly(EGDMA) display similar FT-IR peaks with $[V_2C_4(mim)_2]Br_2$ and EGDMA, respectively. Figure 1B shows Raman spectra of poly($[V_2C_4(mim)_2]Br_2$) and poly(EGDMA). The peaks at 2933 and 2955 cm^{-1} from Figure 1B were due to CH_2 stretching vibration [34].

Curve a shows the characteristic vibration bands (1429, 1343, 1022 cm^{-1}) corresponding to the imidazolium cation [34], respectively. The peaks at 1726, 1461 cm^{-1} are ascribed to C=O, CH_2 bonds of poly(EGDMA) (Curve b) [35]. The results show the successful synthesis of poly($[V_2C_4(mim)_2]Br_2$) and poly(EGDMA). Figure 2 displays the TGA curves of poly($[V_2C_4(mim)_2]Br_2$) and poly(EGDMA). A total weight loss of poly($[V_2C_4(mim)_2]Br_2$) and poly(EGDMA) are 88.9% and 94.4% at 800 °C, respectively. The results suggests the thermal instability of poly($[V_2C_4(mim)_2]Br_2$) and poly(EGDMA). Figure 3 presents the SEM image of poly($[V_2C_4(mim)_2]Br_2$), revealing spherical morphology with diameter ranging from 0.65 to 1.85 μm.

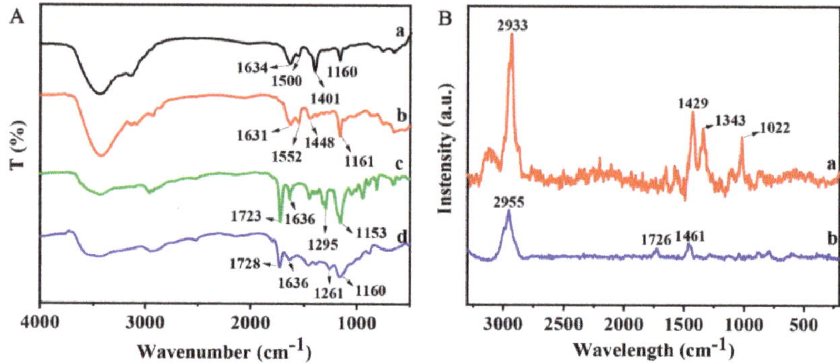

Figure 1. (**A**) FT-IR spectra of $[V_2C_4(mim)_2]Br_2$ (a), poly($[V_2C_4(mim)_2]Br_2$) (b), EGDMA (c), and poly(EGDMA) (d). (**B**) Raman spectra of poly($[V_2C_4(mim)_2]Br_2$) (a) and poly(EGDMA) (b).

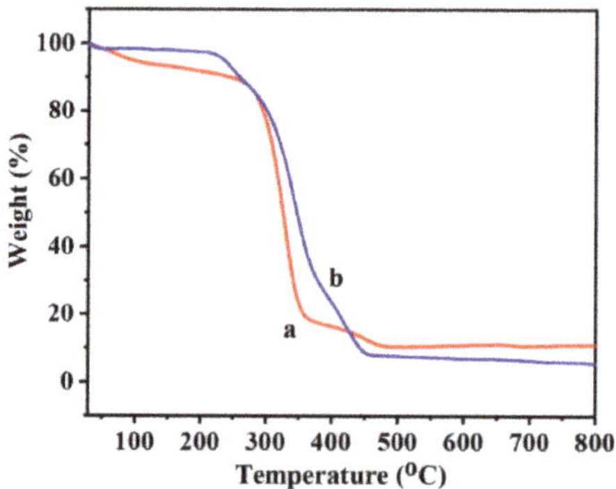

Figure 2. TGA curves of poly($[V_2C_4(mim)_2]Br_2$) (a) and poly(EGDMA) (b).

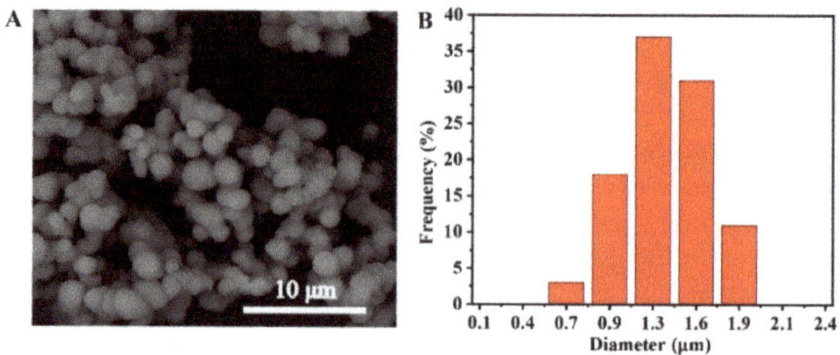

Figure 3. (**A**) SEM image of poly($[V_2C_4(mim)_2]Br_2$). (**B**) Size distribution diagram of poly($[V_2C_4(mim)_2]Br_2$).

3.2. Electrochemical Behavior of Poly([$V_2C_4(mim)_2$]Br_2)

We studied the electrochemical behaviors of the bare GCE and poly([$V_2C_4(mim)_2$]Br_2)/GCE using cyclic voltammetry (CV). Figure 4A shows the CVs of the bare GCE and poly([$V_2C_4(mim)_2$]Br_2)/GCE in 0.1 M KCl containing 5.0 mM $[Fe(CN)_6]^{3-/4-}$. Compared with the bare GCE (curve a), poly([$V_2C_4(mim)_2$]Br_2)/GCE exhibits higher peak current response, suggesting good electrical conductivity of poly([$V_2C_4(mim)_2$]Br_2), which improves the electron transfer kinetics of $[Fe(CN)_6]^{3-/4-}$ at the PIL electrode. In addition, we studied the electrochemical impedance spectroscopy (EIS) of the two electrodes. The charge transfer resistance (Rct) represents the electron transfer kinetics of a redox probe at the electrode surface. We found lower Rct value of poly([$V_2C_4(mim)_2$]Br_2)/GCE when compared with the bare GCE and attributed the results to the good electron transfer conductivity of poly([$V_2C_4(mim)_2$]Br_2).

The effective surface areas of the bare GCE and poly([$V_2C_4(mim)_2$]Br_2)/GCE were calculated from the plots of Q vs. $t^{1/2}$, obtained by chronocoulometry (Figure 4C) in 0.5 mM $K_3[Fe(CN)_6]$ following the Anson method [36,37]. We obtained the effective surface areas of the bare GCE and poly([$V_2C_4(mim)_2$]Br_2)/GCE as 0.0669, and 0.151 cm^2 from the slopes of Q vs. $t^{1/2}$ plot (Figure 4D). The results demonstrate the ability of poly([$V_2C_4(mim)_2$]Br_2) to increase the effective surface area. Therefore, we can infer that the effective surface area of poly([$V_2C_4(mim)_2$]Br_2)/GCE can enhance the total adsorption ability for 4-NP, thus increasing the detection sensitivity.

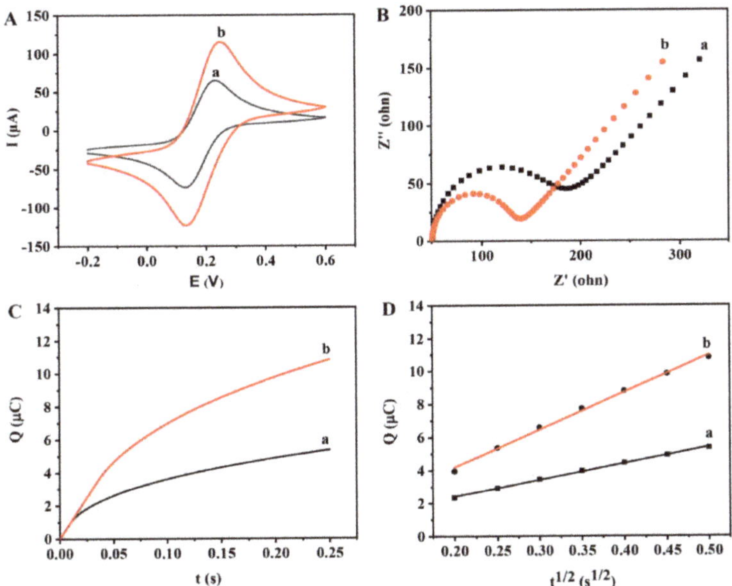

Figure 4. (**A**) Cyclic voltammograms of bare GCE (a) and poly([$V_2C_4(mim)_2$]Br_2)/GCE (b) in 0.1 M KCl containing 5.0 mM $[Fe(CN)_6]^{3-/4-}$. (**B**) Nyquist plots of bare GCE (a) and poly([$V_2C_4(mim)_2$]Br_2)/GCE (b) in 0.1 M KCl containing 5.0 mM $[Fe(CN)_6]^{3-/4-}$. (**C**) Plot of bare GCE (a) and poly([$V_2C_4(mim)_2$]Br_2)/GCE (b) in 5.0 mM $[Fe(CN)_6]^{3-/4-}$. (**D**) Plot of Q-$t^{1/2}$ curves for bare GCE (a) and poly([$V_2C_4(mim)_2$]Br_2)/GCE (b) in 5.0 mM $[Fe(CN)_6]^{3-/4-}$.

To highlight the advantage of PIL, we compared the electrochemical responses of poly([$V_2C_4(mim)_2$]Br_2)/GCE with poly(EGDMA)/GCE. Poly([$V_2C_4(mim)_2$]Br_2)/GCE and poly(EGDMA)/GCE were incubated with a 4-NP solution in phosphate buffer for 6 min, before carrying out the CV and DPV measurements (Figure 5). We noted oxidation of 4-NP on two electrodes (Figure 5A), but no cathodic signal was observed in the scanned potential range. The result indicates an irreversible oxidation process for 4-NP [38]. As shown in Figure 5, poly([V_2C_4(mim)$_2$]Br_2)/GCE exhibited higher CV and DPV peak current responses for 4-NP than those of poly(EGDMA)/GCE. The peak current response for 4-NP of poly([V_2C_4(mim)$_2$]Br_2)/GCE was 3.6 times than that of poly(EGDMA)/GCE. We ascribed the higher current response to the excellent electrical conductivity and large effective surface area of poly([V_2C_4(mim)$_2$]Br_2)/GCE.

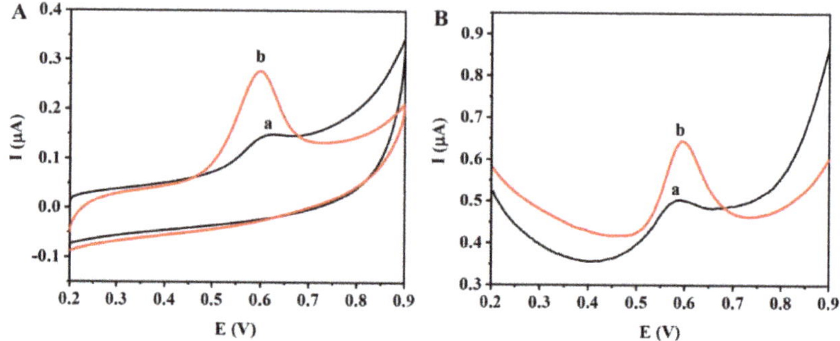

Figure 5. (**A**) Cyclic voltammograms of poly(EGDMA)/GCE (a) and poly([V_2C_4(mim)$_2$]Br_2)/GCE (b) after adsorbing 2 μM 4-NP in 0.1 M phosphate buffer (pH = 6.0) for 6 min. Scan rate: 100 mV/s. (**B**) Differential pulse voltammograms of poly(EGDMA)/GCE (a) and poly([V_2C_4(mim)$_2$]Br_2)/GCE (b) after adsorbing 2 μM 4-NP for 6 min in 0.1 M phosphate buffer (pH = 6.0).

3.3. Optimization of Experimental Conditions

To investigate the influence of pH of the test solution on the current of 4-NP, we measured the DPV of 2 μM 4-NP at the poly([V_2C_4(mim)$_2$]Br_2)/GCE over the pH range of 4–8. As shown in Figure 6A, the peak current increased with pH from 4.0 to 6.0; however, we observed a decrease in the peak current beyond pH 6. Hence, we selected pH 6.0 as the optimum pH for subsequent experiments.

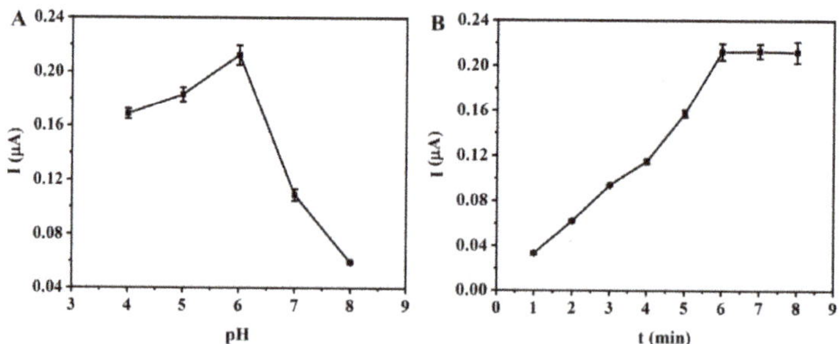

Figure 6. The influence of test solution pH (**A**) and incubation time (**B**) on DPV peak currents at poly([V_2C_4(mim)$_2$]Br_2)/GCE after incubating in 2 μM 4-NP in phosphate buffer.

Moreover, Figure 6B shows increase in the peak current response from until 6 min before reaching a plateau. This indicates that the poly([V_2C_4(mim)$_2$]Br_2)/GCE was saturated by 4-NP adsorption. Thus, further accumulation of 4-NP onto the electrode (beyond 6 min) did not contribute to the enhancement of peak current. Based on our findings, we employed an optimum accumulation time of 6 min for 4-NP adsorption onto the poly([V_2C_4(mim)$_2$]Br_2)/GCE in all subsequent experiments.

3.4. Analytical Performance of Poly([V_2C_4(mim)$_2$]Br_2)/GCE

Under the optimized experimental conditions, the DPV curves of 4-NP at different concentrations were measured by poly([V_2C_4(mim)$_2$]Br_2)/GCE (Figure 7A). We observed a linear increase in the peak current with 4-NP concentration from 0.05 to 5 μM (Figure 7B). We obtained a limit of detection (LOD) of 0.01 μM (S/N = 3) for 4-NP using the following linear regression equation: I(μA) = 0.102 C(μM) + 0.0087 (R^2 = 0.9980). Compared with the other electrodes reported for 4-NP [1,27,39–43], the poly([V_2C_4(mim)$_2$]Br_2)/GCE exhibited satisfactory analytical parameters (Table 1). To demonstrate the advantage of poly([V_2C_4(mim)$_2$]Br_2) sensor, we also investigated the linearity of a traditional cross-linker-based poly(EGDMA) sensor for 4-NP (Figure 7B). The poly(EGDMA)/GCE had a detection limit of 0.8 μM (S/N = 3) for a linear 4-NP concentration range of 1.0 to 5 μM. The above findings clearly demonstrate higher sensitivity of poly([V_2C_4(mim)$_2$]Br_2) based sensor towards 4-NP when compared with the poly(EGDMA)-based sensor.

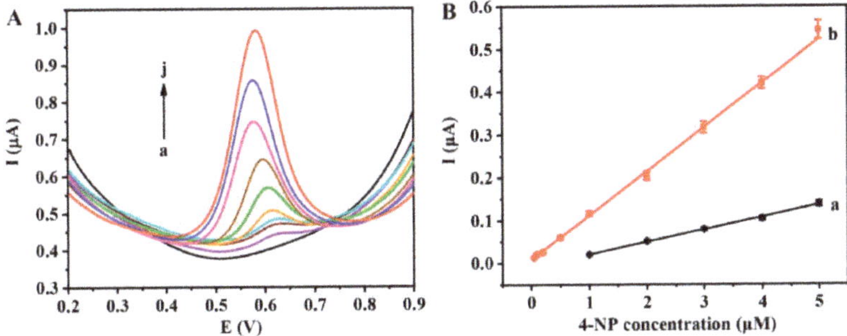

Figure 7. (**A**) Differential pulse voltammograms of poly([V_2C_4(mim)$_2$]Br_2)/GCE for 4-NP from 0 (a), 0.05 (b), 0.1 (c), 0.2 (d), 0.5 (e), 1 (f), 2 (g), 3 (h), 4 (i), 5 (j) μM. (**B**) Calibration plots for 4-NP with poly(EGDMA)/GCE (a) and poly([V_2C_4(mim)$_2$]Br_2)/GCE (b) (*n* = 3).

Table 1. Comparison with other electrochemical methods for the determination of 4-NP.

Detection Method	Linear Range (μM)	Detection Limit (μM)	Ref.
CTAB [a]/carbon paste electrode	0.1–25	0.01	[1]
AuNPs/PILs [b]/GCE	0.1–120	0.033	[27]
β-CD-SH-GR [c]/Au electrode	0.07–70	0.061	[39]
GR [d]-DNA/GCE	0.05–4	0.01	[40]
MIL-101(Cr)@Rgo [e]/GCE	0.1–12.5	0.033	[41]
MIP [f]/AuNPs/TiO_2/GCE	0.95–480	0.32	[42]
IL-FGNS [g]/GCE	0.5–200	0.058	[43]
Poly([V_2C_4(mim)$_2$]Br_2)/GCE	0.05–5	0.01	This work

[a]: Cetyltrimethylammonium bromide; [b]: Poly(1-butyl imidazole-3-(2-ethyl methacrylate) tetrafluoroborate) ionic liquid hollow nanospheres; [c]: Thiol-β-cyclodextrin (β-CD-SH) and graphene (GR) hybrid; [d]: Reduced graphene; [e]: Chromium terephthalate metal-organic frameworks@reduced graphene oxide; [f]: Molecularly imprinted polymer; [g]: Ionic liquid-functionalized grapheme nanosheet.

3.5. *Reproducibility, Stability and Selectivity of Poly([V_2C_4(mim)$_2$]Br_2)/GCE*

To study reproducibility and stability of poly([V_2C_4(mim)$_2$]Br_2)/GCE, we performed DPV measurements of 2 μM 4-NP solution using seven modified GCEs prepared independently. We obtained the relative standard deviation (RSD) as 4.35%, confirming good reproducibility of the method. To check the stability, we carried out similar measurements with the poly([V_2C_4(mim)$_2$]Br_2)/GCE after storing for two weeks at 4 °C, and found that the sensor could retain 93.7% of its original response (2 μM 4-NP), thus exhibiting good stability.

To evaluate the selectivity, we subjected our poly([V_2C_4(mim)$_2$]Br_2)/GCE to detect 4-NP in presence of some possible interfering substances. Table 2 lists the tolerable ratio, which is defined as the maximum amount of foreign species resulting in an error lower than ±5% for detecting 2 μM 4-NP. We found that 100-fold of Na^+, K^+, Fe^{2+}, Mg^{2+}, Ni^{2+}, Co^{2+}, Cu^{2+}, Cl^-, NO_3^-, SO_4^{2-}, and 10-fold hydroquinone, catechol, benzene, 1,2-dimethylbenzene, 2-nitroaniline, 3-nitroaniline and 4-nitroaniline had no influence on the determination of 2 μM 4-NP by the poly([$V_2C_4(mim)_2$]Br_2)/GCE.

Table 2. Effects of foreign species on the determination of 4-NP.

Foreign Species	Tolerable Ratio
Na^+, K^+, Fe^{2+}, Mg^{2+}, Ni^{2+}, Co^{2+}, Cu^{2+}, Cl^-, NO_3^-, SO_4^{2-}	100
Hydroquinone, catechol, benzene, 1,2-dimethylbenzene, 2-nitroaniline, 3-nitroanilie, 4-nitroaniline	10

3.6. Analytical Application

Furthermore, we evaluated the validity of the proposed method using the lake water samples collected from Nanhu Lake and rain water samples in Jiaxing city. The environmental water samples were filtered through 0.45 μm filter before use. Since no 4-NP was detected in the water samples by the poly([$V_2C_4(mim)_2$]Br_2)/GCE, we applied the standard addition method of spiking the samples with 4-NP for the method validation purpose. As shown in Table 3, the recovery was from 97.3% to 104.0%, and the RSD was less than 4.47%. This suggests the potential of the poly([$V_2C_4(mim)_2$]Br_2)/GCE for the detecting 4-NP in environmental water samples.

Table 3. Detection of 4-NP in water samples by poly([$V_2C_4(mim)_2$]Br_2)/GCE (n = 3).

Sample	Added 4-NP (μM)	Found 4-NP (μM)	Recovery (%)	RSD (%)
Lake water	0.1	0.104	104.0	4.38
	1	0.991	99.1	4.47
	2	1.946	97.3	3.45
Rain water	0.1	0.0982	98.2	3.24
	1	1.031	103.1	4.43
	2	2.026	101.3	4.04

4. Conclusions

In an attempt to construct a sensitive electrochemical sensor for 4-NP, we synthesized a cross-linker-based PIL of poly([$V_2C_4(mim)_2$]Br_2) as the sensing agent via a simple one-step free-radical polymerization method. Poly([$V_2C_4(mim)_2$]Br_2)/GCE with a large effective surface area exhibited good electrical conductivity and higher peak current response for 4-NP than that of poly(EGDMA)/GCE, indicating the advantage of poly([$V_2C_4(mim)_2$]Br_2). In addition to excellent sensitivity, the poly([$V_2C_4(mim)_2$]Br_2) based sensor also demonstrated 4-NP detection in environmental water samples.

Author Contributions: J.H., Y.Z., L.L., and R.C. designed the experiments; J.H. performed the experiments; J.H., H.D., Y.Z., Y.Y., H.W., X.Z., L.L., G.Z., and L.G. analyzed the data; J.H., Y.Z., and L.L. wrote the paper; and all authors reviewed, edited, and approved the manuscript.

Funding: This work was supported by the National Natural Science Foundation of China (No. 21507041, 21677060), the Zhejiang Provincial Natural Science Foundation of China under Grant No. LY16B050007 and LQ19B050002, the Open Project of MOE Key Laboratory for Analytical Science of Food Safety and Biology (No. FS18011), and the Program for Science and Technology of Jiaxing (No. 2018AY11002, 2017AY33034).

Conflicts of Interest: The authors declare no conflict of interest.

References

1. Lu, Q.; Zhang, W.; Wang, Z.; Yu, G.; Yuan, Y.; Zhou, Y. A facile electrochemical sensor for nonylphenol determination based on the enhancement effect of cetyltrimethylammonium bromide. *Sensors* **2013**, *13*, 758–768. [CrossRef]

2. Soares, A.; Guieysse, B.; Jefferson, B.; Cartmell, E.; Lester, J. Nonylphenol in the environment: A critical review on occurrence, fate, toxicity and treatment in wastewaters. *Environ. Int.* **2008**, *34*, 1033–1049. [CrossRef]
3. Preuss, T.G.; Gehrhardt, J.; Schirmer, K.; Coors, A.; Rubach, M.; Russ, A.; Jones, P.D.; Giesy, J.P.; Ratte, H.T. Nonylphenol isomers differ in estrogenic activity. *Environ. Sci. Technol.* **2006**, *40*, 5147–5153. [CrossRef] [PubMed]
4. Céspedes, R.; Skryjová, K.; Raková, M.; Zeravik, J.; Fránek, M.; Lacorte, S.; Barceló, D. Validation of an enzyme-linked immunosorbent assay (ELISA) for the determination of 4-nonylphenol and octylphenol in surface water samples by LC-ESI-MS. *Talanta* **2006**, *70*, 745–751. [CrossRef]
5. Lopes, D.; Dias, A.N.; Merib, J.; Carasek, E. Hollow-fiber renewal liquid membrane extraction coupled with 96-well plate system as innovative high-throughput configuration for the determination of endocrine disrupting compounds by high-performance liquid chromatography-fluorescence and diode array detection. *Anal. Chim. Acta* **2018**, *1040*, 33–40.
6. Pastor-Belda, M.; Viñas, P.; Campillo, N.; Hernández-Córdoba, M. Magnetic solid phase extraction with $CoFe_2O_4$/oleic acid nanoparticles coupled to gas chromatography-mass spectrometry for the determination of alkylphenols in baby foods. *Food. Chem.* **2017**, *221*, 76–81. [CrossRef] [PubMed]
7. Han, S.; Li, X.; Wang, Y.; Su, C. A core–shell Fe_3O_4 nanoparticle–CdTe quantum dot–molecularly imprinted polymer composite for recognition and separation of 4-nonylphenol. *Anal. Methods* **2014**, *6*, 2855–2861. [CrossRef]
8. Ai, J.; Guo, H.; Xue, R.; Wang, X.; Lei, X.; Yang, W. A self-probing, gate-controlled, molecularly imprinted electrochemical sensor for ultrasensitive determination of p-nonylphenol. *Electrochem. Commun.* **2018**, *89*, 1–5. [CrossRef]
9. Qin, J.; Guo, J.; Xu, Q.; Zheng, Z.; Mao, H.; Yan, F. Synthesis of pyrrolidinium-type poly(ionic liquid) membranes for antibacterial applications. *ACS Appl. Mater. Int.* **2017**, *9*, 10504–10511. [CrossRef] [PubMed]
10. Zhou, D.; Liu, R.; Zhang, J.; Qi, X.; He, Y.; Li, B.; Yang, Q.; Hu, Y.; Kang, F. In situ synthesis of hierarchical poly(ionic liquid)-based solid electrolytes for high-safety lithium-ion and sodium-ion batteries. *Nano Energy* **2017**, *33*, 45–54. [CrossRef]
11. Osada, I.; de Vries, H.; Scrosati, B.; Passerini, S. Ionic-Liquid-Based Polymer Electrolytes for Battery Applications. *Angew. Chem. Int. Ed.* **2016**, *55*, 500–513. [CrossRef]
12. Qian, W.; Texter, J.; Yan, F. Frontiers in poly(ionic liquid)s: Syntheses and applications. *Chem. Soc. Rev.* **2017**, *46*, 1124–1159. [CrossRef]
13. Wang, P.; Wang, T.; Lin, W.; Lin, H.; Lee, M.; Yang, C. Crosslinked polymer ionic liquid/ionic liquid blends prepared by photopolymerization as solid-state electrolytes in supercapacitors. *Nanomaterials* **2018**, *8*, 225. [CrossRef] [PubMed]
14. Gao, C.J.; Chen, G.J.; Wang, X.C.; Li, J.; Zhou, Y.; Wang, J. A hierarchical meso-macroporous poly(ionic liquid) monolith derived from a single soft template. *Chem. Commun.* **2015**, *51*, 4969–4972. [CrossRef]
15. Suo, X.; Xia, L.; Yang, Q.; Zhang, Z.; Bao, Z.; Ren, Q.; Yang, Y.; Xing, H. Synthesis of anion-functionalized mesoporous poly(ionic liquid)s via a microphase separation-hypercrosslinking strategy: Highly efficient adsorbents for bioactive molecules. *J. Mater. Chem. A* **2017**, *5*, 14114–14123. [CrossRef]
16. Wu, M.; Wang, L.; Zeng, B.; Zhao, F. Ionic liquid polymer functionalized carbon nanotubes-doped poly(3,4-ethylenedioxythiophene) for highly-efficient solid-phase microextraction of carbamate pesticides. *J. Chromatogr. A* **2016**, *1444*, 42–49. [CrossRef] [PubMed]
17. Wang, R.; Yuan, Y.; Yang, X.; Han, Y.; Yan, H. Polymethacrylate microparticles covalently functionalized with an ionic liquid for solid-phase extraction of fluoroquinolone antibiotics. *Microchim. Acta* **2015**, *182*, 2201–2208. [CrossRef]
18. Chen, L.; Mei, M.; Huang, X.; Yuan, D. Sensitive determination of estrogens in environmental waters treated with polymeric ionic liquid-based stir cake sorptive extraction and liquid chromatographic analysis. *Talanta* **2016**, *152*, 98–104. [CrossRef] [PubMed]
19. Liu, C.; Liao, Y.; Huang, X. Fabrication of polymeric ionic liquid-modified magnetic adsorbent for extraction of apolar and polar pollutants in complicated samples. *Talanta* **2017**, *172*, 23–30. [CrossRef]
20. Liu, C.; Deng, Q.; Fang, G.; Dang, M.; Wang, S. Capillary electrochromatography immunoassay for alpha-fetoprotein based on poly(guanidinium ionic liquid) monolithic material. *Anal. Biochem.* **2017**, *530*, 50–56. [CrossRef]

21. Wang, Q.; Yun, Y. Nonenzymatic sensor for hydrogen peroxide based on the electrodeposition of silver nanoparticles on poly(ionic liquid)-stabilized graphene sheets. *Microchim. Acta* **2013**, *180*, 261–268. [CrossRef]
22. Sánchez-Paniagua López, M.; López-Ruiz, B. Electrochemical biosensor based on ionic liquid polymeric microparticles. An analytical platform for catechol. *Microchem. J.* **2018**, *138*, 173–179. [CrossRef]
23. Cui, K.; Lu, X.; Cui, W.; Wu, J.; Chen, X.; Lu, Q. Fluorescent nanoparticles assembled from a poly(ionic liquid) for selective sensing of copper ions. *Chem. Commun.* **2011**, *47*, 920–922. [CrossRef] [PubMed]
24. Wang, Y.; Li, C.; Wu, T.; Ye, X. Polymerized ionic liquid functionalized graphene oxide nanosheets as a sensitive platform for bisphenol A sensing. *Carbon* **2018**, *129*, 21–28. [CrossRef]
25. Yu, Y.; Yu, C.; Yin, T.; Ou, S.; Sun, X.; Wen, X.; Zhang, L.; Tang, D.; Yin, X. Functionalized poly (ionic liquid) as the support to construct a ratiometric electrochemical biosensor for the selective determination of copper ions in AD rats. *Biosens. Bioelectron.* **2017**, *87*, 278–284. [CrossRef]
26. Li, J.; Li, Q.; Zeng, Y.; Tang, T.; Pan, Y.; Li, L. An electrochemical sensor for the sensitive determination of phenylethanolamine A based on a novel composite of reduced graphene oxide and poly (ionic liquid). *RSC Adv.* **2015**, *5*, 717–725. [CrossRef]
27. Ren, J.; Gu, J.; Tao, L.; Yao, M.; Yang, X.; Yang, W. A novel electrochemical sensor of 4-nonylphenol based on a poly(ionic liquid) hollow nanosphere/gold nanoparticle composite modified glassy carbon electrode. *Anal. Methods* **2015**, *7*, 8094–8099. [CrossRef]
28. Ding, S.; Hu, X.; Guan, P.; Zhang, N.; Li, J.; Gao, X.; Zhang, X.; Ding, X.; Du, C. Preparation of surface-imprinted microspheres using ionic liquids as novel cross-linker for recognizing an immunostimulating peptide. *J. Mater. Sci.* **2017**, *52*, 8027–8040. [CrossRef]
29. Ma, W.W.; Row, K.H. Solid-phase extraction of chlorophenols in seawater using a magnetic ionic liquid molecularly imprinted polymer with incorporated silicon dioxide as a sorbent. *J. Chromatogr. A* **2018**, *1559*, 78–85. [CrossRef] [PubMed]
30. Zhu, X.; Zeng, Y.; Zhang, Z.; Yang, Y.; Zhai, Y.; Wang, H.; Liu, L.; Hu, J.; Li, L. A new composite of graphene and molecularly imprinted polymer based on ionic liquids as functional monomer and cross-linker for electrochemical sensing 6-benzylaminopurine. *Biosens. Bioelectron.* **2018**, *108*, 38–45. [CrossRef] [PubMed]
31. Yuan, J.; Antonietti, M. Poly (ionic liquid) latexes prepared by dispersion polymerization of ionic liquid monomers. *Macromolecules* **2011**, *44*, 744–750. [CrossRef]
32. Zhang, Q.; Wu, S.; Zhang, L.; Lu, J.; Verproot, F.; Liu, Y.; Xing, Z.; Li, J.; Song, X. Fabrication of polymeric ionic liquid/graphene nanocomposite for glucose oxidase immobilization and direct electrochemistry. *Biosens. Bioelectron.* **2011**, *26*, 2632–2637. [CrossRef]
33. Mao, Y.; Bao, Y.; Gan, S.; Li, F.; Niu, L. Electrochemical sensor for dopamine based on a novel graphene-molecular imprinted polymers composite recognition element. *Biosens. Bioelectron.* **2011**, *28*, 291–297. [CrossRef]
34. Ana-Maria, G.; Lucian, R.; Mihaela, B.; Ioan, B.; Camelia, B. Sensitive detection of endocrine disrupters using ionic liquid–single walled carbon nanotubes modified screen-printed based biosensors. *Talanta* **2011**, *85*, 2007–2013.
35. Filipecka, K.; MiedzińSki, R.; Sitarz, M.; Filipecki, J.; Makowska-Janusik, M. Optical and vibrational properties of phosphorylcholine-based contact lenses-experimental and theoretical investigations. *Spectrochim. Acta A* **2017**, *176*, 83–90. [CrossRef] [PubMed]
36. Anson, F.C. Application of Potentiostatic Current Integration to the Study of the Adsorption of Cobalt(III)-(Ethylenedinitrilo(tetraacetate) on Mercury Electrodes. *Anal. Chem.* **1964**, *36*, 932–934. [CrossRef]
37. Li, S.; Lei, S.; Yu, Q.; Zou, L.; Ye, B. A novel electrochemical sensor for detecting hyperin with a nanocomposite of ZrO_2-SDS-SWCNTs as decoration. *Talanta* **2018**, *185*, 453–460. [CrossRef] [PubMed]
38. Su, D.; Zhang, Y.; Wang, Z.; Wan, Q.; Yang, N. Decoration of graphene nano platelets with gold nanoparticles for voltammetry of 4-nonylphenol. *Carbon* **2017**, *117*, 313–321. [CrossRef]
39. Xue, F.; Gao, Z.; Sun, X.; Yang, Z.; Yi, L.; Chen, W. Electrochemical determination of environmental hormone nonylphenol based on composite film modified gold electrode. *J. Electrochem. Soc.* **2015**, *162*, H338–H344. [CrossRef]
40. Zeng, L.; Zhang, A.; Zhu, X.; Zhang, C.; Liang, Y.; Nan, J. Electrochemical determination of nonylphenol using differential pulse voltammetry based on a graphene–DNA-modified glassy carbon electrode. *J. Electroanal. Chem.* **2013**, *703*, 153–157. [CrossRef]

41. Zhang, Y.; Yan, P.; Wan, Q.; Yang, N. Integration of chromium terephthalate metal-organic frameworks with reduced graphene oxide for voltammetry of 4-nonylphenol. *Carbon* **2018**, *134*, 540–547. [CrossRef]
42. Huang, J.; Zhang, X.; Liu, S.; Lin, Q.; He, X.; Xing, X.; Lian, W.; Tang, D. Development of molecularly imprinted electrochemical sensor with titanium oxide and gold nanomaterials enhanced technique for determination of 4-nonylphenol. *Sens. Actuators B-Chem.* **2011**, *152*, 292–298. [CrossRef]
43. Meng, X.; Yin, H.; Xu, M.; Ai, S.; Zhu, J. Electrochemical determination of nonylphenol based on ionic liquid-functionalized graphene nanosheet modified glassy carbon electrode and its interaction with DNA. *J. Solid State Electr.* **2012**, *16*, 2837–2843. [CrossRef]

Article

Construction of Novel Polymerizable Ionic Liquid Microemulsions and the In Situ Synthesis of Poly(Ionic Liquid) Adsorbents

Aili Wang [1,2,*], Shuhui Li [1], Hou Chen [1,*], Ying Liu [1] and Xiong Peng [2]

1 School of Chemistry and Material Science, Ludong University, Yantai 264025, China; lsh5354@163.com (S.L.); 15563833132@163.com (Y.L.)
2 Guangdong Provincial Key Lab of Green Chemical Product Technology, Guangzhou 510640, China; cepeng.xiong@mail.scut.edu.cn
* Correspondence: wang.aili@mail.scut.edu.cn (A.W.); chenhou@ldu.edu.cn (H.C.); Tel.: +86-0535-6697921 (A.W.); +86-0535-6697933 (H.C.); Fax: +86-0535-6697921 (A.W.); +86-0535-6697933 (H.C.)

Received: 22 January 2019; Accepted: 12 March 2019; Published: 18 March 2019

Abstract: This paper reports the successful construction of novel polymerizable ionic liquid microemulsions and the in situ synthesis of poly(ionic liquid) adsorbents for the removal of Zn^{2+} from aqueous solution. Dynamic light-scattering data were used to confirm the polymerization media and to illustrate the effect of the crosslinker dosage on the droplet size of the microemulsion. FTIR and thermal analysis were employed to confirm the successful preparation of the designed polymers and characterize their thermostability and glass transition-temperature value. The optimization of the adsorption process indicates that the initial concentration of Zn^{2+}, pH, adsorbent dosage and contact time affected the adsorption performance of poly(ionic liquid)s toward Zn^{2+}. Furthermore, our research revealed that the adsorption process can be effectively described by the pseudo second-order kinetic model and the Freundlich isotherm model.

Keywords: polymerizable ionic liquid microemulsions; poly(ionic liquid)s; adsorption

---

1. Introduction

Poly(ionic liquid)s (PILs), which possess ionic liquid species in each of the repeating units, have attracted much attention in the field of polyelectrolyte preparation, porous membrane synthesis, adsorption and separation [1,2]. Among these applications, the adsorption of organic pollutants and metal ions in industrial wastewater has received increased attention from researchers [3]. PILs are polymerized from ionic liquid monomers, and thus possess the excellent physical and chemical properties of ionic liquids, such as molecular designability, negligible evaporation, and excellent thermal stability [1,4–6]. Moreover, PILs have overcome most of the disadvantages of ionic liquids, such as the high viscosity, recycling difficulty, and consequent toxic and corrosive risk [3,7].

Earlier research on PILs focused on their adsorption performance towards organics with ionic groups or aromatic nucleus, such as anionic dyes and triazole fungicides [8,9]. Currently, PILs are also showing an increased potential for heavy metal ion adsorption [10]. Interestingly, the "designability" of the ionic liquid monomers makes hydrophobic anions inducible into the molecular chain of PILs to introduce hydrophobic polymers, which increases their reusability for adsorbing heavy metal ions. Moreover, the adsorptive selectivity could be enhanced due to the "designability" of the monomers. The reported application of PILs focused on the removal of Cr(VI), but the adsorption capacity was unsatisfied. The hydrophobic PILs synthesized by Kong only showed a maximum adsorption capacity of 17.9 mg/g for hexavalent chromium [10]. The sulfonic acid-functionalized PILs synthesized by

Khiratkar showed a higher adsorption capacity for hexavalent chromium in water, but there was still a gap compared to other adsorbents [11].

The traditional synthesis method of PILs was almost bulk-free radical polymerization at normal pressure. Our previous research showed that functional materials could be synthesized in situ within ionic liquid microemulsions [12]. Nowadays, microemulsions involving ionic liquids have attracted much attention in the world. As a "green solvent", ionic liquid has been applied in various areas depending on its low volatility, designability, negligible vapor pressure and excellent stability advantages [13]. Previous research indicated that ionic liquids could replace any of the constituents in traditional microemulsions and form a new system named ionic liquid microemulsion [1,3,14,15]. This innovative finding not only enlarged the research scale of ionic liquid and microemulsion to a greater extent, but also promoted the development of the relevant subjects. Due to the controllability of the nanosized structure of the ionic liquid microemulsion droplets, nanomaterials such as TiO_2, phosphate nanocrystals and polyaniline could be easily synthesized by controlling the microstructure of the designed microemulsions [16–18]. Moreover, ionic liquid monomers could also be used as the polar and/or non-polar phase of microemulsions, and these types of polymeric ionic liquid microemulsions could be used as an in situ synthesis media for PILs [19].

In the present study, we constructed a new type of ionic liquid microemulsion, in which all components were polymerizable. The polar phase was a mixture of 1-allyl-3-methylimidazolium chloride ([AMIM][Cl]) and cross-linking agent poly(ethylene glycol) diacrylate (PEGDA). The nonpolar phase was 1-allyl-3-methylimidazolium hexafluorophosphate ([AMIM][PF_6]). Sodium 3-(allyloxy)-2-hydroxypropane-1-sulfonate (HAPS) was used as the surfactant, and no co-surfactant was involved. Compressed bulk free-radical polymerization was adopted to synthesize PILs in the polymerizable ionic liquid microemulsions (PILMs). The resulting PILs were used as an adsorbent for metal ions in aqueous solution. Zn(II), which mainly comes from metallurgical processing, electroplating and batteries industries [20], was selected here as the object metal ion. The effect of PEGDA dosage on the droplet size and size distribution of PILMs was investigated by dynamic light scattering (DLS). The influence of thermosynthesis time, synthesis temperature, adsorbent dosage, pH, initial concentration and contact time on the characteristic and adsorption capability of PILs for Zn(II) removal were investigated in detail.

2. Materials and Methods

2.1. Materials

[AMIM][PF_6] (>99 wt%) and [AMIM][Cl] (>99 wt%) were purchased from Lanzhou Institute of Chemical Physics. HAPS (>99 wt%) was obtained from Hanerchem Ltd. PEGDA (Mn = 575), azobisisobutyronitrile (AIBN, >99 wt%) and zinc chloride ($ZnCl_2$, >98 wt%) were purchased from Sigma Aldrich. Prior to the experiment, HAPS was vacuum-dried at 70 °C for 6 h to remove excess water. The other chemicals were used without further purification.

2.2. Methods

In a typical experiment, the polymerization process of PILs included the following steps: (1) [AMIM][PF_6], [AMIM][Cl], HAPS and PEGDA at different mass ratios were mixed at 60 °C for 40 min under moderate magnetic stirring, and the microemulsions named PILM-1 to PILM-4 were obtained. (2) The mixture was transferred into a sealed Teflon-lined autoclave and subsequently kept at 80–120 °C for 12–28 h. (3) The reaction set was cooled to room temperature, and the obtained products were precipitated by ethyl acetate followed by deionized water three times. (4) The brown precipitate was collected by rotary evaporation and vacuum-dried overnight at 40 °C, and the final products, named PIL-1 to PIL-5, were obtained. The general recipe used in the synthesis of PILs is shown in Table 1. In addition, the synthesis process of the PILs is shown in Scheme 1.

Table 1. The general recipe used in the preparation of poly(ionic liquid)s (PILs [a]).

Sample Names	$m_{(AMIMCl)}$ /g	$m_{(AMIMPF6)}$ /g	$m_{(HAPS)}$ /g	$m_{(PEGDA)}$ /g	$m_{(AIBN)}$ /g	Temp. /°C
PILM-1	3.0611	4.0031	6.0065	0.7221	\	60
PILM-2	3.0618	4.0243	5.9999	1.0731	\	60
PILM-3	3.0481	4.0031	6.0145	1.7653	\	60
PILM-4	3.0814	4.0005	6.0062	2.8480	\	60
PIL-1	3.0615	4.0207	6.0125	2.8482	1.1188	80
PIL-2	3.0588	4.0515	6.0003	2.8548	1.1196	90
PIL-3	3.0492	4.0054	6.0082	2.8395	1.1172	100
PIL-4	3.0630	4.0943	6.0168	2.8704	1.1361	110
PIL-5	3.0346	4.0721	6.0270	2.8555	1.1231	120

[a] The uncertainty limits are ±0.1%.

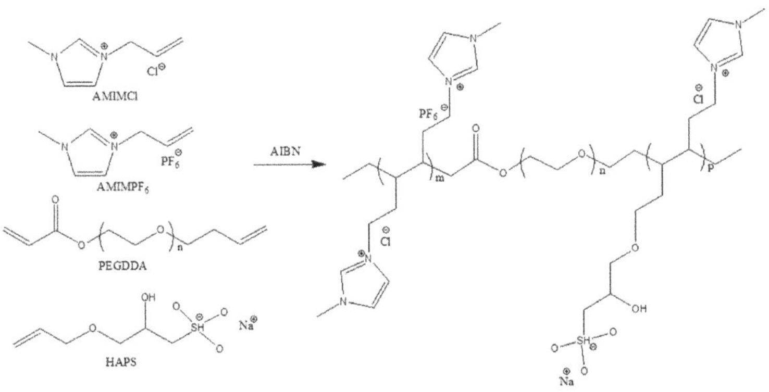

Scheme 1. The synthesis process of PILs.

2.3. Characterization of PILs

A Malvern Nano ZS was applied to determine the droplet size, size distribution and zeta potential of PILMs. A Perkin Elmer Spectrum 2000 was applied to determine the FTIR spectrum of PILMs. The thermogravimetric data of the PILMs were collected by a Netzsch STA449C F3 instrument at a heating rate of 10 °C/min under argon atmosphere from room temperature to 600 °C. Differential scanning calorimeter (DSC) analysis was performed on a Netzsch 204F1 calorimeter and carried out over a temperature range of −50 °C to 150 °C with a heating rate of 10 °C/min. The yield was obtained using the gravimetric method. The equilibrium degree of swelling (E_s) of the PILs was measured by gravimetry.

E_s is defined as:

$$E_s = \frac{W_s}{W_d},$$

where W_s is the weight of the PILs after equilibrium swelling, and W_d is the dry weight of the PILs.

In order to evaluate the PILs' adsorption capacities toward Zn(II) in aqueous solution, the batch adsorption experiments were performed as follows. In a typical experiment, the mixtures of the designed PILs and 20 mL of 50 mg/L Zn(II) solution were added into a series of 100 mL Erlenmeyer flask with a glass stopper. At a constant temperature of 25 ± 1 °C in a thermostatic water bath, the samples were shaken at about 100 rpm for certain contact time. The Zn(II) ion concentrations of the after-adsorbed samples were determined by flame atomic absorption spectroscopy (AAS). The adsorption capacities (q_e in mg/g) of the designed PILs were calculated according to Equation (1):

$$q_e = (C_0 - C_e) \times \frac{V}{m}, \quad (1)$$

where C_0 is the initial concentrations of Zn(II) ion (mg/L), C_e is the equilibrium concentrations of Zn(II) ion (mg/L), V is the volume of Zn(II) solution (L), and m is the dosage of PILs (g).

3. Results and Discussion

3.1. Droplet Size and Size Distribution

Transparent PILMs involving [AMIM][Cl], [AMIM][PF_6], PEGDA and HAPS were constructed successfully. Normally, dynamic light scattering (DLS) is used to determine the droplet size and size distribution of microemulsions containing ionic liquids [21]. Herein, the average droplet size and size distribution of the [AMIM][Cl]/[AMIM][PF_6]/PEGDA/HAPS microemulsions with different amounts of PEGDA were collected and shown in Figure 1. It can be noticed that the size distribution of each PILMs was monodispersed, and the average size ranged from 10–40 nm, which was in accordance with the microemulsion's nature (<150 nm) [22]. Besides, with the increasing amount of PEGDA, the average size of PILMs increased slightly, and this could be attributed to the droplet structure of PILMs. According to the molar weight of each component in the construction of the PILMs, PEGDA was stable as the inner phase of the microemulsion when the ionic liquids acted as the outer phase. Therefore, the increment dosage of PEGDA enlarged the size of the inner core, and hence, increased the droplet size. The zeta potential values of the [AMIM][Cl]/[AMIM][PF_6]/PEGDA/HAPS microemulsions were -24.13 ± 0.85 mV indicating that the polymerizable ionic liquid microemulsions had good stability.

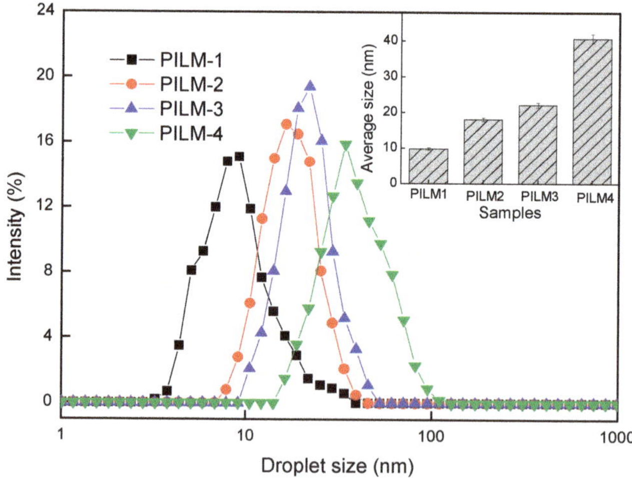

Figure 1. Size and size distribution of the [AMIM][Cl]/[AMIM][PF_6]/PEGDA/HAPS microemulsion with different amounts of PEGDA.

3.2. FTIR Analysis

The in situ synthesized PILs with different reaction temperatures—80 °C, 90 °C, 100 °C, 110 °C and 120 °C, represented by PIL-1, PIL-2, PIL-3, PIL-4 and PIL-5, respectively—were characterized by Fourier transform infrared spectroscopy (FTIR). The results are shown in Figure 2, where the adsorption peak around 1724 cm^{-1} and 1090 cm^{-1} can be attributed to the stretching vibration of

C=O and C–O in PEGDA, respectively. The peak that appeared at 2913 cm^{-1} was attributed to the C–H stretching vibration in the saturated carbon chain. The absorption peak located at 830 cm^{-1} and 550 cm^{-1} was characteristic of PF_6^- and Cl^-, respectively, which suggested the successful synthesis of the PILs within the PILMs.

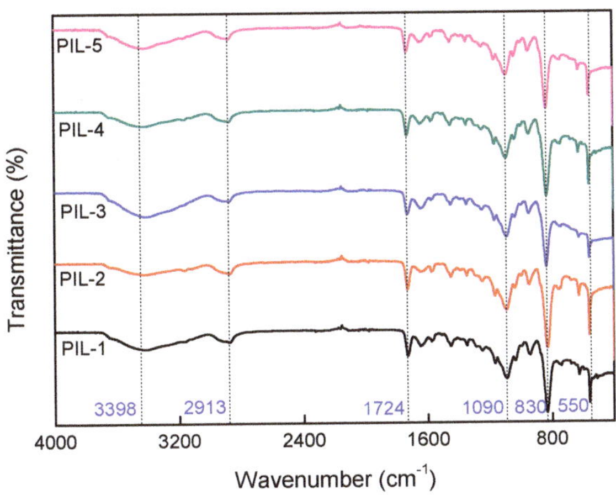

Figure 2. FTIR spectra of PILs synthesized under different temperature.

3.3. Thermal Analysis of PILs

The thermogravimetry curves of [AMIM][Cl], [AMIM][PF_6] and the as-prepared PILs are presented in Figure 3A. Each PIL weight loss exhibited two distinct temperature scopes. Due to the evaporation of the absorbed water, the PILs showed an initial stage of weight loss in the range of room temperature to 100 °C. This absorbed water was mainly combined with chlorine salt because of the water-absorbing quality of [AMIM][Cl]. The second stage of weight loss was found in the range of 300 °C to 400 °C, which can be ascribed to the fracture of the imidazole ring in the PILs as well as the [AMIM][Cl] and [AMIM][PF_6] monomers. These results showed that PILs synthesized within the PILMs had combined the thermogravimetry characteristics of [AMIM][Cl] and [AMIM][PF_6]. The water resistance of PILs synthesized over 80 °C manifested improvements to a certain extent compared to [AMIM][Cl]. PIL-3, which was synthesized at 100 °C, showed the best thermogravimetry performance within the range of study.

The swelling behavior of PILs in Zn(II) solutions was further investigated with the gravimetry method. As shown in Table 2, the designed PILs could be well-swelled in solutions, and the complexation of the ester group of PEGDA with Zn(II), and the hydroxy and the sulfonic group of HAPS with Zn(II) occurred simultaneously. Since five PILs had the same ionic liquid and crosslinker content, the effect of the synthesis temperature of the PILs on E_s was not obvious.

Table 2. E_s of PILs in water.

PIL	PIL-1	PIL-2	PIL-3	PIL-4	PIL-5
E_s	21.37	21.01	20.82	21.18	20.85

Figure 3B shows the DSC curves of different PILs. The results showed that the glass transition temperatures (T_g) of the PILs were approximately −30 °C, which indicated their phase transformation point. Altering the reaction temperature of the PILs did not produce obvious changes in the T_g position.

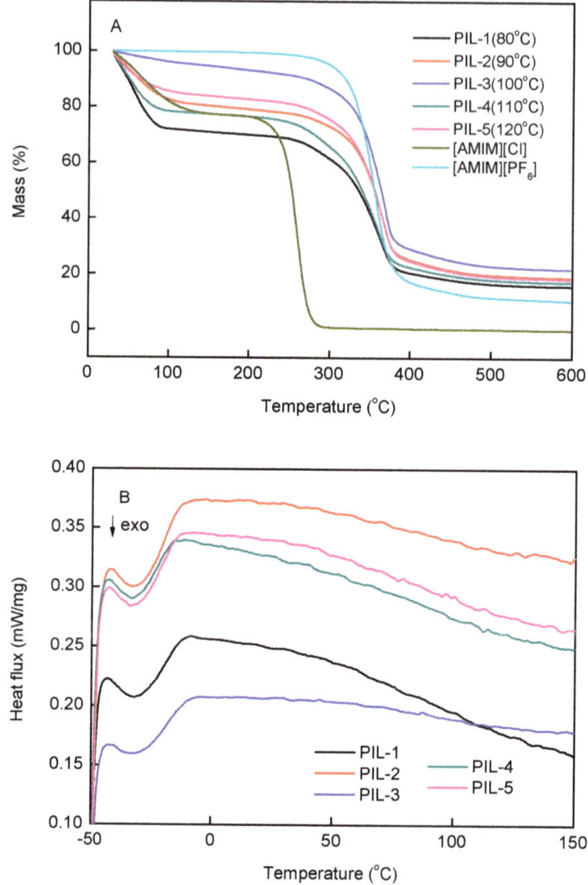

Figure 3. Thermogravimetric curves of [AMIM][Cl], [AMIM][PF_6] and PILs. (**A**) DSC curves of PILs (**B**).

3.4. Influence on the Adsorption Potential of PILs

The as-prepared PILs were employed in the removal of Zn^{2+} from water, and PIL-3 was selected as the model adsorbent because of its distinct properties mentioned above. In addition, the adsorption performance of PIL-5 was investigated for comparison (volume—25 mL, pH—4.5, dose—10 mg, time—8 h and temperature—25 ± 1 °C). As shown in Figure 4A, within the studied range of the initial concentration of Zn^{2+} (C_0), the PIL-3 sample exhibited better adsorption performance than PIL-5. Moreover, with the increment of C_0, the adsorption capacities of both PIL samples toward Zn^{2+} increased gradually.

The effects of pH and adsorbent dosage on the adsorption of Zn^{2+} onto PIL-3 were investigated, and the results are provided in Figure 4B,C, respectively. It can be observed that the saturated adsorption capacity grew up slightly with the increase in solution pH from 3.5 to 5.5, suggesting that electrostatic attraction played a role in the adsorption process (C_0—0.01 M, volume—25 mL, dose—10 mg, time—8 h and temperature—25 ± 1 °C). Notably, lowering the adsorbent dosage brings about the increasing of adsorption capacity, which can be ascribed to the excess adsorption sites (C_0—0.01 M, volume—25 mL, pH—4.5, time—8 h and temperature—25 ± 1 °C).

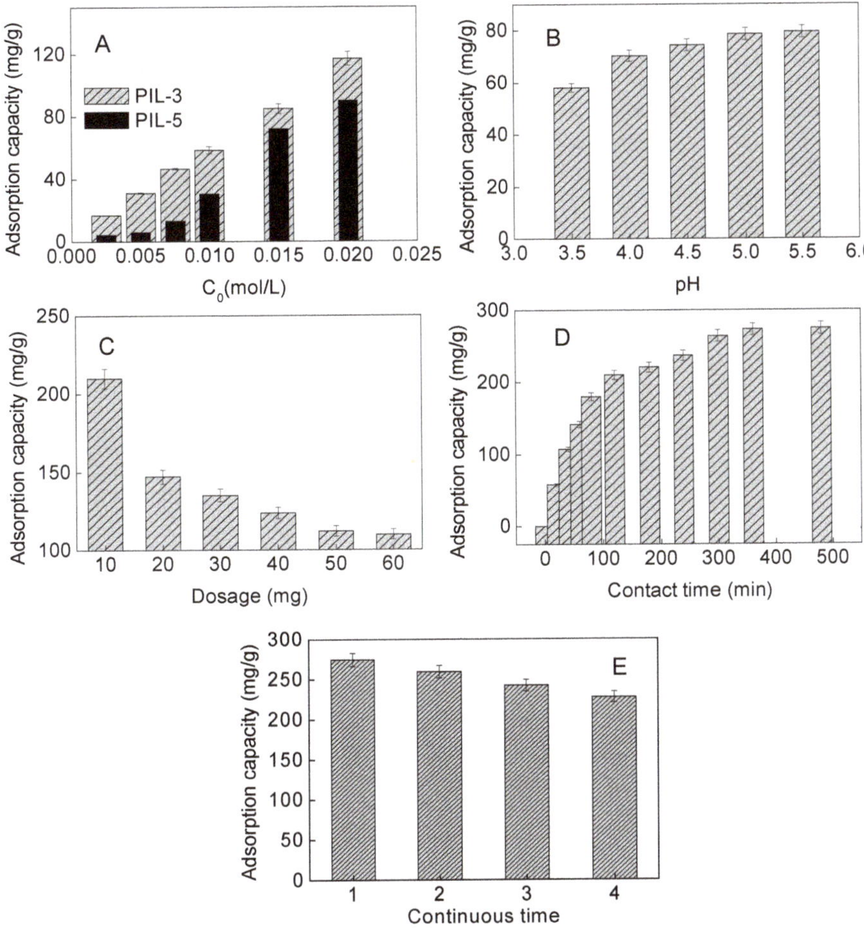

Figure 4. The effect of C_0 on the adsorption performance of PIL-3 and PIL-5 for Zn^{2+} (**A**), the effect of pH (**B**), adsorbent dosage (**C**), contact time (**D**) and continuous times (**E**) on the adsorption performance of PIL-3 for Zn^{2+}.

Figure 4D shows the adsorption behavior of PIL-3 with Zn^{2+} as a function of contact time (C_0—0.01 M, volume—25 mL, pH—4.5, dose—10 mg and temperature—25 ± 1 °C). The adsorption capacity of PIL-3 toward Zn^{2+} increased rapidly from 58.34 to 179.81 mg/g when the contact time was raised from 20 to 80 min. On further raising of the contact time to 300 min, the adsorption capacity increased slowly. In order to achieve sufficient adsorption, 480 min was selected as the optimum contact time. The influence on the adsorption potential of the designed PILs indicated that the maximum adsorption capacity of the PILs for Zn(II) was greater than the values of various reported adsorbents [23–28].

In addition, PIL-3 was easily released in 0.1 mol/L HCl after absorbing Zn^{2+} from the water. After washing with deionized water and drying in a vacuum at 50 °C, PIL-3 was able to be continuously used more than three times without a significant decrease in Zn^{2+} removal efficiency (<12%), as shown in Figure 4E.

3.5. Adsorption Kinetics

The sorption kinetics for the adsorption of Zn^{2+} onto PIL-3 was investigated, and the resulting data were fitted in the pseudo-first-order and pseudo-second-order kinetic models, as shown in Figure 5. The calculated kinetic parameters are shown in Table 3. Obviously, the adsorption process was more aligned with the pseudo-second-order model (R^2 = 0.997) than with the pseudo-first-order model (R^2 = 0.874). Besides, for the pseudo-second-order model, the calculated saturated adsorption capacity value was also closer to the experimental data. These findings indicate that the ion exchange and physical adsorption on the surface and the hydroxyl complexation may control the valence forces between PIL-3 and Zn^{2+}.

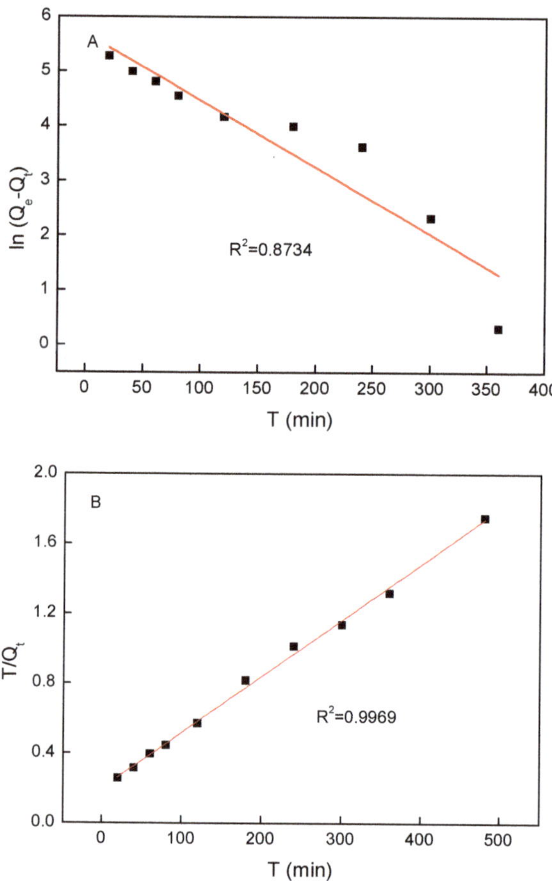

Figure 5. Pseudo-first-order kinetic model fitting (**A**) and pseudo-second-order kinetic model fitting (**B**) of the adsorption process for PIL-3 toward Zn^{2+}.

Table 3. Adsorption kinetic parameters of Zn^{2+} on PIL-3.

$Q_{e,exp}$ mg/g	Pseudo-First Order			Pseudo-Second Order			
	$Q_{e,cal}$ mg/g	k_1	R^2	$Q_{e,cal}$ mg/g	$k_2 \times 10^4$	H mg/(g min)	R^2
273.03	196.17	0.0120	0.8743	312.5	0.5172	5.0505	0.9969

3.6. Adsorption Isotherms

The adsorption isotherm of Zn^{2+} on PIL-1 was investigated using typical Langmuir and Freundlich adsorption isotherm models [29,30]. The fitting of the two models is presented in Figure 6, and the calculated parameters are listed in Table 4. It can be noted that the adsorption process was more effectively described by the Freundlich adsorption isotherm model (R^2 = 0.9970) than the Langmuir adsorption isotherm model (R^2 = 0.1056). This indicates that the adsorption surface of PIL-1 was heterogeneous. The reason for this could be the increasing porosity of PILs during the thermosynthesis process, which conformed to the Freundlich assumption [31]. In addition, the value of n (1.1325) showed that the adsorption of Zn^{2+} onto PIL-3 was favorable.

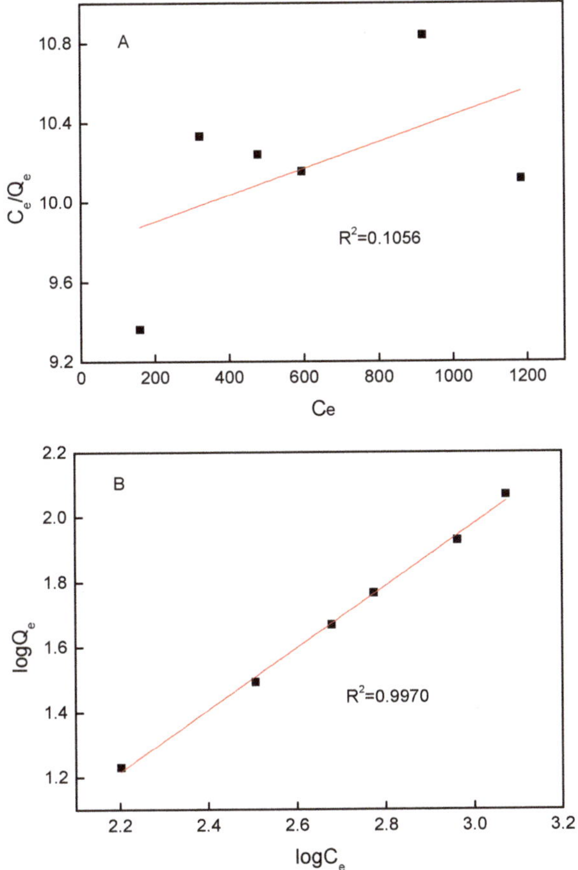

Figure 6. Langmuir (**A**) and Freundlich (**B**) adsorption isotherm for the adsorption of Zn^{2+} onto PIL-3.

Table 4. Adsorption isotherm parameters of Zn^{2+} on PIL-3.

Langmuir Isotherm			Freundlich Isotherm		
Q_{max} mg/g	k_L L/mg	R^2	K_f L/g	n	R^2
0.1505	0.6801	0.1056	8.9991	1.1325	0.9970

4. Conclusions

Novel PILMs were constructed and used as the in situ synthesis media for the polymerization of PILs. The designed polymers were proved to be useful adsorbents for the removal of Zn^{2+} from aqueous solution. DLS analysis showed the ionic liquid microemulsion nature of the polymerization media and indicated that the dosage of PEGDA affected the average size of the PILMs. The FTIR spectra confirmed the successful preparation of the designed PILs. Thermal analysis showed that the PILs synthesized within PILMs combined the thermogravimetric characteristics of [AMIM][Cl] and [AMIM][PF_6], showing good thermostability, and the T_g of PILs were about −30 °C. Within the range of study, PIL-3 showed the best thermogravimetric performance. C_0, pH, adsorbent dosage and contact time affected the adsorption performance of Zn^{2+} on PILs. In addition, the adsorption process was effectively described by the pseudo-second-order kinetic model and the Freundlich isotherm model.

Author Contributions: Conceptualization, A.W.; methodology, A.W. and H.C.; validation, A.W.; investigation, S.L. and Y.L.; resources, A.W.; data curation, S.L. and Y.L.; writing—original draft preparation, S.L. and Y.L.; writing—review and editing, A.W.; funding acquisition, A.W. and H.C.

Funding: This research was supported by the Research Fund Program of Guangdong Provincial Key Lab of Green Chemical Product Technology (GC201820), China Postdoctoral Science Foundation funded project (2018 M630947), the School Scientific Research Foundation for Introduced Talent (LA2016001), and the Key Program for Basic Research of Natural Science Foundation of Shandong Province (ZR2018ZC0946).

Acknowledgments: The authors would gratefully acknowledge the support from the Guangdong Provincial Key Lab of Green Chemical Product Technology.

Conflicts of Interest: The authors declare no conflict of interest.

References

1. Yuan, J.Y.; Mecerreyes, D.; Antonietti, M. Poly(ionic liquid)s: An update. *Prog. Polym. Sci.* **2013**, *38*, 1009–1036. [CrossRef]
2. Mecerreyes, D. Polymeric ionic liquids: Broadening the properties and applications of polyelectrolytes. *Prog. Polym. Sci.* **2011**, *36*, 1629–1648. [CrossRef]
3. Wang, S.J.; Ma, H.L.; Peng, J.; Zhang, Y.W.; Chen, J.; Wang, L.L.; Xu, L.; Li, J.Q.; Zhai, M.L. Facile synthesis of a novel polymeric ionic liquid gel and its excellent performance for hexavalent chromium removal. *Dalton Trans.* **2015**, *44*, 7618–7625. [CrossRef] [PubMed]
4. Patrascu, C.; Gauffre, F.; Nallet, F.; Bordes, R.; Oberdisse, J.; de Lauth-Viguerie, N.; Mingotaud, C. Micelles in ionic liquids: Aggregation behavior of alkyl poly(ethyleneglycol)-ethers in 1-butyl-3-methyl-imidazolium type ionic liquids. *ChemPhysChem* **2006**, *7*, 99–101. [CrossRef] [PubMed]
5. Yuan, J.Y.; Antonietti, M. Poly(ionic liquid)s: Polymers expanding classical property profiles. *Polymer* **2011**, *52*, 1469–1482. [CrossRef]
6. Qian, W.J.; Texter, J.; Yan, F. Frontiers in poly(ionic liquid)s: Syntheses and applications. *Chem. Soc. Rev.* **2017**, *46*, 1124–1159. [CrossRef] [PubMed]
7. Mudraboyina, B.P.; Obadia, M.M.; Allaoua, I.; Sood, R.; Serghei, A.; Drockenmuller, E. 1,2,3-Triazolium-Based Poly(ionic liquid)s with Enhanced Ion Conducting Properties Obtained through a Click Chemistry Polyaddition Strategy. *Chem. Mater.* **2014**, *26*, 1720–1726. [CrossRef]
8. Zhao, W.F.; Tang, Y.S.; Xi, J.; Kong, J. Functionalized graphene sheets with poly(ionic liquid)s and high adsorption capacity of anionic dyes. *Appl. Surf. Sci.* **2015**, *326*, 276–284. [CrossRef]
9. Liu, C.; Liao, Y.M.; Huang, X.J. Extraction of triazole fungicides in environmental waters utilizing poly (ionic liquid)-functionalized magnetic adsorbent. *J. Chromatogr. A* **2017**, *1524*, 13–20. [CrossRef]
10. Mi, H.; Jiang, Z.G.; Kong, J. Hydrophobic Poly(ionic liquid) for Highly Effective Separation of Methyl Blue and Chromium Ions from Water. *Polymers* **2013**, *5*, 1203–1214. [CrossRef]
11. Khiratkar, A.G.; Kumar, S.S.; Bhagat, P.R. Designing a sulphonic acid functionalized benzimidazolium based poly(ionic liquid) for efficient adsorption of hexavalent chromium. *RSC Adv.* **2016**, *6*, 37757–37764. [CrossRef]
12. Wang, A.L.; Chen, L.; Zhang, J.X.; Sun, W.C.; Guo, P.; Ren, C.Y. Ionic liquid microemulsion-assisted synthesis and improved photocatalytic activity of $ZnIn_2S_4$. *J. Mater. Sci.* **2017**, *52*, 2413–2421. [CrossRef]

13. Hallett, J.P.; Welton, T. Room-Temperature Ionic Liquids: Solvents for Synthesis and Catalysis. 2. *Chem. Rev.* **2011**, *111*, 3508–3576. [CrossRef]
14. Lee, S.; Yim, T.; Park, Y.D.; Mun, J. Room Temperature Ionic Liquid-Activated Nafion Polymer Electrolyte for High Temperature Operation. *Polym.-Korea* **2018**, *42*, 682–686. [CrossRef]
15. Wang, A.L.; Chen, L.; Jiang, D.Y.; Zeng, H.Y.; Yan, Z.C. Vegetable oil-based ionic liquid microemulsion biolubricants: Effect of integrated surfactants. *Ind. Crops Prod.* **2014**, *62*, 515–521. [CrossRef]
16. Mirhoseini, F.; Salabat, A. Ionic liquid based microemulsion method for the fabrication of poly(methyl methacrylate)-TiO_2 nanocomposite as a highly efficient visible light photocatalyst. *RSC Adv.* **2015**, *5*, 12536–12545. [CrossRef]
17. Zhang, C.; Chen, J.; Zhu, X.; Zhou, Y.; Li, D. Synthesis of Tributylphosphate Capped Luminescent Rare Earth Phosphate Nanocrystals in an Ionic Liquid Microemulsion. *Chem. Mater.* **2009**, *21*, 3570–3575. [CrossRef]
18. He, D.L.; Xia, S.B.; Zhou, Z.; Zhong, J.F.; Guo, Y.N.; Yang, R.H. Electropolymerization of polyaniline in ionic liquid (bmim PF6)/water microemulsion. *J. Exp. Nanosci.* **2013**, *8*, 103–112. [CrossRef]
19. Wang, A.; Li, S.; Zhang, L.; Chen, H.; Li, Y.; Hu, L.; Peng, X. Ionic liquid microemulsion-mediated in situ thermosynthesis of poly(ionic liquid)s and their adsorption properties for Zn(II). *Polym. Eng. Sci.* **2019**. [CrossRef]
20. Ennigrou, D.J.; Ali, M.B.; Dhahbi, M. Copper and Zinc removal from aqueous solutions by polyacrylic acid assisted-ultrafiltration. *Desalination* **2014**, *343*, 82–87. [CrossRef]
21. Warisnoicharoen, W.; Lansley, A.B.; Lawrence, M.J. Nonionic oil-in-water microemulsions: The effect of oil type on phase behaviour. *Int. J. Pharm.* **2000**, *198*, 7–27. [CrossRef]
22. Zhang, M.; Wang, Y.Y.; Bai, T.C. Phase Diagrams, Density, and Viscosity for the Pseudoternary System of {Propan-2-yl Tetradecanoate (IPM) (1) plus Tween 80 (21) plus Propan-1-ol (22) (2) plus Water (3)}. *J. Chem. Eng. Data* **2012**, *57*, 2023–2029. [CrossRef]
23. Krishnan, K.A.; Sreejalekshmi, K.G.; Vimexen, V.; Dev, V.V. Evaluation of adsorption properties of sulphurised activated carbon for the effective and economically viable removal of Zn(II) from aqueous solutions. *Ecotoxicol. Environ. Saf.* **2016**, *124*, 418–425. [CrossRef] [PubMed]
24. Musso, T.B.; Parolo, M.E.; Pettinari, G.; Francisca, E.M. Cu(II) and Zn(II) adsorption capacity of three different clay liner materials. *J. Environ. Manag.* **2014**, *146*, 50–58. [CrossRef] [PubMed]
25. Mousavi, S.J.; Parvini, M.; Ghorbani, M. Experimental design data for the zinc ions adsorption based on mesoporous modified chitosan using central composite design method. *Carbohydr. Polym.* **2018**, *188*, 197–212. [CrossRef]
26. Nakanishi, K.; Tomita, M.; Kato, K. Synthesis of amino-functionalized mesoporous silica sheets and their application for metal ion capture. *J. Asian Ceram. Soc.* **2015**, *3*, 70–76. [CrossRef]
27. Bao, S.Y.; Tang, L.H.; Li, K.; Ning, P.; Peng, J.H.; Guo, H.B.; Zhu, T.T.; Liu, Y. Highly selective removal of Zn(II) ion from hot-dip galvanizing pickling waste with amino-functionalized Fe_3O_4@SiO_2 magnetic nano-adsorbent. *J. Colloid Interface Sci.* **2016**, *462*, 235–242. [CrossRef]
28. Monier, M.; Abdel-Latif, D.A. Preparation of cross-linked magnetic chitosan-phenylthiourea resin for adsorption of Hg(II), Cd(II) and Zn(II) ions from aqueous solutions. *J. Hazard. Mater.* **2012**, *209*, 240–249. [CrossRef]
29. Banerjee, S.S.; Chen, D.H. Fast removal of copper ions by gum arabic modified magnetic nano-adsorbent. *J. Hazard. Mater.* **2007**, *147*, 792–799. [CrossRef]
30. Da'na, E.; Sayari, A. Adsorption of copper on amine-functionalized SBA-15 prepared by co-condensation: Equilibrium properties. *Chem. Eng. J.* **2011**, *166*, 445–453. [CrossRef]
31. Oke, I.A.; Olarinoye, N.O.; Adewusi, S.R.A. Adsorption kinetics for arsenic removal from aqueous solutions by untreated powdered eggshell. *Adsorpt.-J. Int. Adsorpt. Soc.* **2008**, *14*, 73–83. [CrossRef]

 *nanomaterials*

Review

New Developments in Material Preparation Using a Combination of Ionic Liquids and Microwave Irradiation

Yannan Wang, Qidong Hou, Meiting Ju and Weizun Li *

College of Environmental Science and Engineering, Nankai University, Tianjin 300350, China; wangyannan@nankai.edu.cn (Y.W.); houqidong@nankai.edu.cn (Q.H.); jumeit@nankai.edu.cn (M.J.)
* Correspondence: liweizun@nankai.edu.cn; Tel.: +86-135-1221-2566

Received: 8 March 2019; Accepted: 16 April 2019; Published: 22 April 2019

Abstract: During recent years, synthetic methods combining microwaves and ionic liquids became accepted as a promising methodology for various materials preparations because of their high efficiency and low energy consumption. Ionic liquids with high polarity are heated rapidly, volumetrically and simultaneously under microwave irradiation. Hence, combination of microwave irradiation as a heating source with ionic liquids with various roles (e.g., solvent, additive, template or reactant) opened a completely new technique in the last twenty years for nanomaterials and polymers preparation for applications in various materials science fields including polymer science. This review summarizes recent developments of some common materials syntheses using microwave-assisted ionic liquid method with a focus on inorganic nanomaterials, polymers, carbon-derived composites and biomass-based composites. After that, the mechanisms involved in microwave-assisted ionic-liquid (MAIL) are discussed briefly. This review also highlights the role of ionic liquids in the reaction and crucial issues that should be addressed in future research involving this synthesis technique.

Keywords: ionic liquids; microwave; synthetic methods; nanomaterials; polymers

1. Introduction

Green chemistry has received increasing attention during the past few decades because of energy crises and environmental pollution. The green chemistry principles led to the development of cleaner and more benign chemical processes, especially in chemical syntheses. Both ionic liquids (ILs) and microwave irradiation are known as a promising technology tool capable of contributing to green chemistry development because of their fast and easy applications, high efficiency and relatively environmental friendliness.

ILs are molten salts in a liquid form at low temperatures (usually below 100 °C) and are composed of large organic cations and small inorganic or organic anions. Comparing with conventional organic solvents, ILs have many unique properties, such as negligible vapor pressure, high temperature stability, non-volatility, chemical stability, large electrochemical stability window and ionic conductivity. Some ILs are also relatively environmentally friendly solvents compared with organic solvents because of their low vapor pressures especially when taking in account volatile organic compounds (VOCs) [1–9].

There are currently three different generations of ILs that have been identified. The first generation of ILs were mainly composed of cations like dialkylimidazolium or alkylpyridinium derivatives and anions like chloroaluminate or other metal halides, which have been described as toxic, non-biodegradable and oxygen-sensitive [10]. After that, the water and oxygen sensitive anions were replaced by halides (Cl^-, Br^-, I^-) or anions such as BF_4^-, PF_6^- and $C_6H_5COO^-$, which are stable in water and air. These ILs possess lower melting points, different solubilities in classic organic solvents, and different viscosities, with an important disadvantage of low toxicity. This generation of ILs attract wide attention and provide interesting and novel application in materials areas [10]. The third generation of ILs named deep eutectic solvents (DESs) based on more hydrophobic and stable anions such as $[(CF_3SO_2)_2N^-]$, sugars, amino or organic acids, alkylsulfates or alkylphosphates, and cations such as choline [11]. DESs are formed by hydrogen bond acceptors and hydrogen bond donors, which can be associated with each other by means of hydrogen bond interactions. Due to their low cost and analogous physico-chemical properties, DESs have attracted considerable attentions in recent years [12–14]. Generally, ILs can be synthesized by chemical synthesis, electrochemical synthesis, microwave-assisted synthesis and ultrasound-accelerated processes [15]. Because of variety of cations and anions, approximately 10^{18} different combinations are possible for IL syntheses [3]. Therefore, ILs will always be referred to as "designed liquids" with properties able to be tuned by adjusting their cations and anions [16–20]. In other words, ILs can be tweaked to meet different requirements by changing types of anions and cations used. For instance, the combination of an imidazolium cation with Cl^- or BF_4^- results in a hydrophilic IL, whereas its combination with PF_6^- results in a hydrophobic IL [17]. Furthermore, ILs containing metal speciations demonstrate additional functionalities such as magnetic, optical or catalytic activities [21,22]. Thus, ILs demonstrate wide applicability to the synthesis of various materials.

On the other hand, consumption of energy for heating is also an integral part of any chemical process. Microwave irradiation is one of the highly desirable energy sources to overcome this problem. Microwave irradiation refers to high frequency electromagnetic waves with wavelengths ranging from 1 mm to 1 m and alternating current signals with frequencies in the 0.3–300 GHz range. Exposure to microwaves results in rapid volumetric heating without the heat conduction process, thus, uniform heating can be achieved in a short period of time [23–28]. In comparison with traditional heating methods, not only can microwave heating rapidly provide high temperature and pressure and significantly decrease reaction time in a closed system but also can replace many organic solvents with water or even enable the reactions to proceed under solvent-free conditions with maximum efficiency [29,30]. Microwave heating provides several additional reaction parameters including power of the microwaves, combination of various organic solvents or even aqueous systems [31,32], inert atmosphere, etc., all of which provide possibilities to obtain a variety of morphologies [33]. Taking advantages of all the above-mentioned benefits, microwaves are widely used not only for common reactions such as materials syntheses, polymerization reactions and biomass extraction but also for some unique process such as hierarchical self-assembly of nanomaterials [32,34], etc.

ILs consist of organic multi-ring and/or organic alkane chains and bulky inorganic anions with high polarizability in a microwave field, which enables them to be heated uniformly and quickly throughout the whole reaction volume. Therefore, ILs are known as excellent microwave absorbers [35]. The first usage of ILs in a microwave-assisted process was reported by Varma and Namboodiri in 2001 [36]. Since then, an ever-increasing number of studies on microwave-assisted ionic-liquid (MAIL) methods were reported. The combination of microwaves and ILs delivers advantages of both, which provides a great potential to meet an ever increasing demand for effective, non- or low-toxicity and economic chemical processes [37].

This paper reviews recent progress on applications of the microwave/ILs technique for rapid and eco-friendly syntheses of functional materials. Preparation of inorganic nanomaterials (such as metal nanoparticles, metal oxides and other complex metal structures using this method) is discussed first. Then, an overview of the polymerization reactions conducted using MAIL methods is provided, followed by a research overview on syntheses of carbon-derived composites. Finally, a review of preparation techniques and various applications of biomass-based nanocomposites and biomass extraction products concludes this paper. The mechanisms and prospects of this method are also discussed briefly.

2. Preparation of Inorganic Nanomaterials

Over the past few decades, inorganic nanomaterials science experienced exponential growth in various fields because of unique physical and chemical properties of nanomaterials [38]. Properties of inorganic materials often depend on their morphology and dimensionality, thus, a lot of strategies to control shape of such nanomaterials were developed and reported in the literature [39]. ILs are widely used for inorganic material syntheses as microwave absorbers, solvents, additives and even as templates to optimize synthesis parameters.

2.1. Metal Nanostructures

The very first report on the preparation of inorganic nano-metals was published by Zhu et al. in 2004 [40]. They synthesized single-crystalline tellurium nanorods and nanowires by MAIL method at 180 °C using a N-butylpyridinium tetrafluoroborate ([BuPy][BF_4]) IL. The whole synthesis took only 10 min. In this work, they demonstrated that combination of an IL and microwave heating played an important role in the synthesis of these metal nanostructures. When conventional heating methods (e.g., oil bath) and other solvent were used, fewer or no nanostructures formed. Other studies also reported successful syntheses of nanostructures of Au, Pt and Rh using the MAIL technique [41–51]. Many researchers emphasize that ILs act as steric stabilizers, solvents and/or microwave absorbers during synthesis of metal nanostructures [41].

Gold nanostructures are the most studied metal nanomaterials because of their wide use. Li et al. synthesized large-size single-crystal gold nanosheets by microwave heating of $HAuCl_4$ in a 1-butyl-3-methylimidazolium tetrafluoroborate ([Bmim][BF_4]) IL for 10 min without any additional template agents [42]. These Au nanosheets were over 30 μm in size. Most recently, Bhawawet et al. designed colloidal gold nanoparticles capped with oleylamine to use them for synthesis of gold nanoparticles by the MAIL method (Figure 1) [43]. They demonstrated shorter reaction times (from 30 s to 1 h) when microwaves were used compared to conventional methods. Their MAIL reactions involved toluene and 1-butyl-1-methylpyrrolidinium bis(trifluoromethylsulfonyl) imide ([C_4mpy][Tf_2N]). Usage of this IL as reaction solvent increased polydispersity of Au nanoparticles (NPs) and reduced reaction time. They also demonstrated that their ILs could be reused and could provide the same reaction effect as fresh ILs.

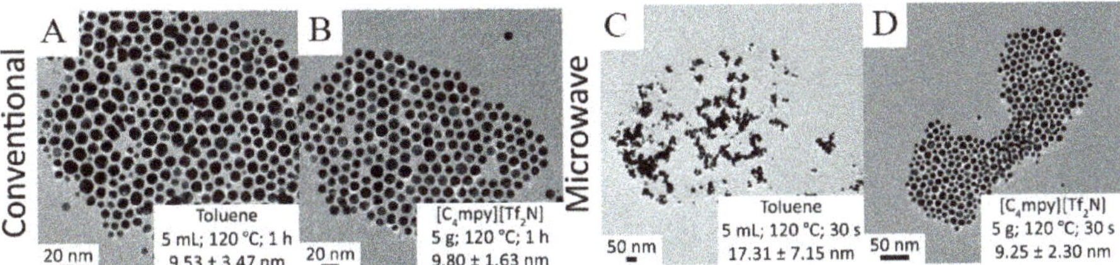

Figure 1. Scheme (**A–D**) representative transmission electron microscopy (TEM) images of Au nanoparticles (NPs) synthesized using conventional and microwave methods carried out in either toluene (panels A and C) or [C_4mpy][Tf_2N] (panels B and D) [43].Copyright Royal Society of Chemistry, 2018.

Li et al. synthesized Ag nanoparticles by a microwave-assisted hydrothermal technique using 1-dodecyl-3-methylimidazolium chloride ([C_{12}mim]Cl) [44]. Their transmission electron microscopy (TEM) showed spherical Ag NPs ~12 nm in size on average. Morphology of Ag NPs was homogeneous. Additionally, particles were crystalline and showed no significant agglomerations. All these factors were attributed to IL participation during Ag NP synthesis. Coulomb coupling forces between Cl^- and [C_{12}mim]$^+$ ions resulted in their alignment, which, in turn, ensured an arrayed structure of the final reaction product. These Ag NPs improved cell activation properties.

The MAIL method was also used for synthesis of other noble metal nanoparticles. For example, Safavi synthesized spherical Pt and Pd nanostructures using various IL media [45]. Schütte et al. prepared Cu, Zn and Cu/Zn brass alloy nanoparticles from metal amidinate precursors using [Bmim][BF_4] [46]. Another study also reported the important role of [Bmim][BF_4] in the synthesis of transition-metal nanoparticles (M-NPs) (M = Ni, Cu). [47].

Carbon materials (e.g., graphene) are well-known substrates for metal NP deposition [48]. Synthesis of metal NPs on such materials can also be implemented using ILs and microwaves. Thus, Lee et al. synthesized Pt NPs on graphene/graphite oxide using 2-hydroxyethanaminium formate ([$HOCH_2CH_2NH_3$][HCO_2]) in combination with microwaves [49] and obtained structures with catalytic hydrogenation capability much higher than commercially-available catalysts. Esteban et al. synthesized stable hybrid iridium@graphene (Ir@TRGO) nanomaterials in a [Bmim][BF_4] IL also using microwave irradiation [50]. Ir@TRGO nanomaterials with average particle diameters <5 nm were used as catalysts of benzene and cyclohexene hydrogenation.

2.2. Metal Oxide Nanostructures

2.2.1. Zinc Oxide (ZnO)

ZnO is a wide bandgap semiconductor with a variety of practical applications. Wang et al. was the first research group to report ZnO synthesis using the MAIL method [51]. They obtained flower- and needle-like ZnO using [Bmim][BF_4] as an additive upon exposure of the whole synthesis system to microwaves for only 5–10 min. They observed that the morphology of the resulting product was dependent on the temperature of the microwaves and IL amount used during the synthesis. Morphology of ZnO changed between flower-like and needle-like by changing the reaction conditions (Figure 2). Flower- and shuttle-like ZnO nanomaterials were later synthesized by the MAIL method using 1-butyl-3-methylimidazolium chloride ([Bmim]Cl) and [Bmim][BF_4] ILs [52]. It was concluded that besides shorter reaction time, movement and polarization of ions under microwave irradiation resulted in the formation of transient and anisotropic micro-domains in the system, which led to the formation of ZnO with different morphologies. Thus, both the IL and microwave irradiation played an important role in the generation of ZnO with different morphologies.

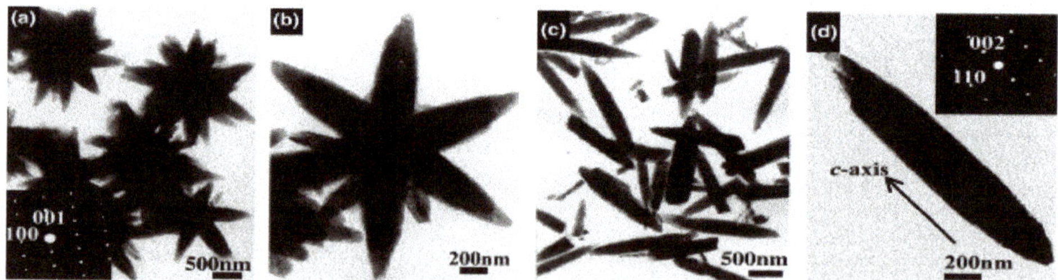

Figure 2. TEM micrographs: (**a**,**b**) Flowerlike ZnO; (**c**,**d**) needlelike ZnO [51]. Copyright Elsevier, 2004.

Because of their favorable bandgap values, ZnO-based materials are usually used as photoanodes and photocatalysts. Rabieh et al. synthesized photocatalytic ZnO NPs using a [Bmim]Cl IL and 400 W microwave irradiation for only 1 min [53]. In this case, the IL acted as a surface agent and as a microwave absorbent. The resulting spherical ZnO NPs showed excellent photocatalytic performance toward photodegradation of the malachite green. Esmaili et al. also prepared nanocrystalline ZnO in a 1-ethyl-3-methylimidazolium ethylsulfate ([Emim][$EtSO_4$]) IL for methylene blue (MB) photodegradation [54]. They proposed that using the IL as a reaction medium could decrease ZnO crystalline size and improve its nucleation rate, which eventually would lead to increased photocatalytic activity. One example using ZnO as a photoanode for solar cells is sensitized ZnO particles prepared with different amounts of a 1,3-dimethylimidazolium iodide ([Dmim]I) IL (da Trindade et al.) [55].

In their system the IL was a dry solid. It did not change ZnO phase formation and/or morphology, but decreased ZnO particle size and converted shallow defects into deep ones, which led to a significant increase of photocurrent density.

2.2.2. Titanium Dioxide (TiO_2)

TiO_2 is most frequently used as a photocatalyst because of its low cost, non-toxicity, high activity and stability. TiO_2-based nanomaterials with different morphologies were fabricated to enhance photocatalytic activity of TiO_2 under visible light. Some studies reported that ILs could facilitate formation of hollow TiO_2 [56], self-assembled TiO_2 nanosheets [57], mesoporous TiO_2 [58], etc.

Recent research reported synthesis of TiO_2 with higher photocatalytic activity using microwave irradiation combined with ILs. Zhang et al. first proposed synthesis of micro-sheet anatase with reactive (001) facets using microwave and a [Bmim][BF_4] IL [59], as shown in Figure 3. The IL in their study provided a fluorine-rich environment and helped to expose (001) facets. Usage of microwaves decreased reaction time and increased crystallinity of TiO_2. As a microwave absorbent, the IL also enhanced microwave efficiency. Chen et al. also mentioned that $[BF_4]^-$ anion in the IL was a critical factor for the synthesis of Ti^{3+} self-doped TiO_2 hollow nanocrystals under microwave irradiation [60]. They used [Bmim][BF_4] to take advantage of its microwave absorbing and morphology-controlling properties. As a result, they obtained TiO_2 mesoporous nanocrystals. It was suggested that the IL also facilitated decreasing recombination of photoelectrons and holes, which normally favors high photocatalytic activity [61].

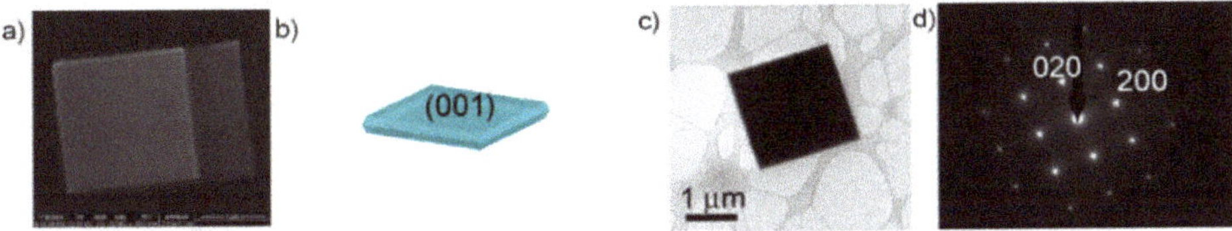

Figure 3. Field emission scanning electron microscope (FESEM) image (**a**), simulated model (**b**), TEM image of a representative anatase single crystal recorded along the (001) axis (**c**), selected area electron diffraction (SAED) pattern and (**d**) high resolution transmission electron microscope (HRTEM) image [59]. Copyright Royal Society of Chemistry, 2009.

Xiao et al. synthesized in situ carbon nanotubes (CNTs) threaded with TiO_2 single crystals with exposed (001) active facets (CNTs–TiO_2) in a 1-methyl imidazolium tetrafluoroborate ([Hmim]BF_4) IL using microwave irradiation [62]. The system was heated at 13 °C min/L to 150 °C and kept at that temperature for 30 min. Field emission scanning electron microscope (FESEM) images of the resulting CNTs–TiO_2 structures with different initial concentrations of $TiCl_3$ are shown in Figure 4. To prepare CNTs–TiO_2-4 (see Figure 4), which demonstrated the highest photocatalytic activity for NO removal, 0.5 M of $TiCl_3$ was used as a precursor. They suggested that presence of $[BF_4]^-$ groups was critical for the formation of a photocatalyst with such high activity. These groups (1) assisted rutile→anatase phase transition, (2) coordinated with Ti^{3+} to oxidize TiO_2 crystals and (3) helped formation of decahedrons with exposed (001) facets. Microwaves helped to increase the surface of the "super-hot" dots to accelerate formation of TiO_2 crystals.

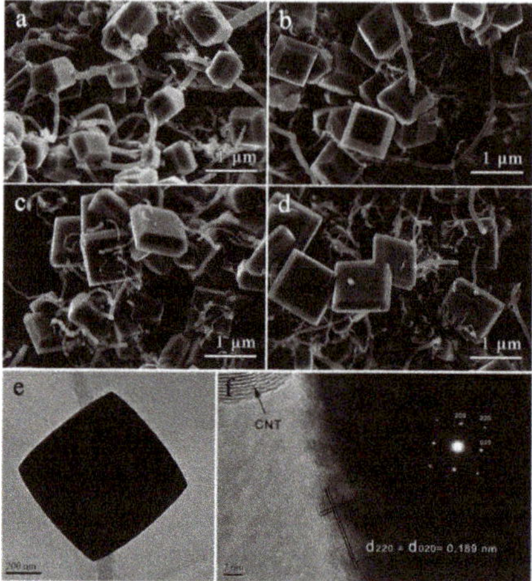

Figure 4. FESEM images of carbon nanotubes (CNTs)–TiO_2 (different concentrations of $TiCl_3$) (**a**–**d**), TEM (**e**) and HRTEM (**f**) image of CNTs–TiO_2-4 (0.5M $TiCl_3$), together with the CNTs–TiO_2-4 SAED image (inset) [62]. Copyright Royal Society of Chemistry, 2016.

2.2.3. Other Metal Oxide Nanomaterials

CuO is a narrow bandgap transition metal oxide widely used for photoconductive and photochemical applications [63]. CuO synthesis by MAIL methods can change the structure of nano-CuO, improve its performance or increase the reaction efficiency. Different morphologies of CuO were prepared by MAIL methods: Nanosheets and nano-whiskers [64], flower- and leaf-like nanomaterials [65], feather- and flower-like crystals [66], CuO NPs [67] and CuO/Cu_2O composites [68].

Flower-like Cu_2O with high photochemical activity also can be prepared by the MAIL method [69] often using 1-octyl-3-methylimidazolium trifluoroacetate ([Omim][TA]) and [Bmim][BF_4] ILs. ILs again played an important role in controlling Cu_2O-based nanomaterials morphologies, which are needed to enhance Cu_2O performance in different applications. Xia et al. achieved different morphologies of CuO crystals (flower-like to leaf-like) by controlling the concentration of an [Omim][TA] IL with high ionic conductivity and polarizability (see Figure 5) [65]. The first step of their morphology-controlling mechanism involved microwave energy absorption by $[Omim]^+$, which led to high heating rate and shorter reaction time. The second step was weak long-range ordering of [Omim][TA] at higher concentrations, which prevented particles accumulating and forming ordered and/or organized structures [70]. Hence, different concentrations of IL resulted in the formation of CuO structures with different morphologies.

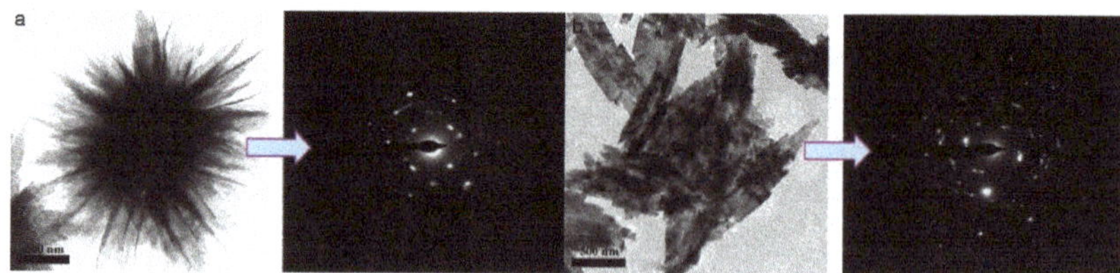

Figure 5. TEM and SAED images of flower-like and leaf-like CuO nanosheets prepared in a reflux system at 80 °C for 10 min under microwave irradiation, (**a**) 2 mL ionic liquid 1-octyl-3-methylimidazolium trifluoroacetate ([Omim]TA) and (**b**) 3 mL ionic liquid [Omim]TA. Scale bar: (**a**) 500 nm and (**b**) 500 nm [65]. Copyright Elsevier, 2009.

Iron oxides are functional materials of great industrial and scientific importance because of their chemical stability and non- or low-toxicity. Ahmed et al. fabricated α-Fe_2O_3 quantum dots using a [Bmim][BF_4] IL and a microwave-assisted solution synthesis method [71]. These quantum dots were ~10–13 nm in diameter with the outstanding photocatalytic properties. Cao and Zhu reported a microwave- and IL-assisted hydrothermal method for the synthesis of α-Fe_2O_3 and Fe_3O_4 NPs with good photocatalytic activity [72]. They demonstrated that it was indeed the [Bmim][BF_4] IL that influenced phase composition and morphology of the final product. Yin et al. showed that temperature and time of microwave exposure also influenced morphology of Fe_3O_4 during MAIL synthesis with a [Bmim][BF_4] IL [73]. Fe_3O_4 nanowires were obtained at 70 °C while nanorods were obtained at 90 °C.

The MAIL method was also successfully used to synthesize other metal oxide nanostructures. Jadhav et al. obtained controlled MgO nanostructures with various green IL solvents using microwave irradiation [74]. Their ILs were N-methyl imidazolium and 3-methyl pyridinium cations with various halide anions. They used ILs not only as solvents but also as structural directing agents. MgO nanostructures with various morphologies were obtained using different ILs and MW conditions (Figure 6). SnO, a very popular anode material, can also be synthesized by the MAIL method. Qin et al. synthesized SnO at low temperature with excellent cycling performance using 1-ethyl-3-methlyl-imidazolium acetate ([Emim][Ac]) and microwave irradiation [75]. ILs played a role of a microwave absorbent to reduce reaction temperature and time.

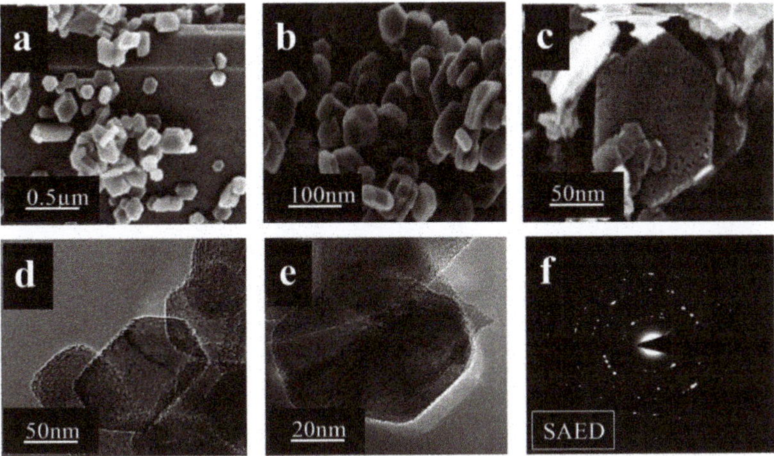

Figure 6. FESEM (**a**–**c**) and FETEM (**d**,**e**) images, and SAED pattern (**f**) of MgO nanohexagons obtained in bis(3-methyl imidazolium-yl) butane dichloride [74]. Copyright Royal Society of Chemistry, 2006.

2.2.4. Other Complex Metal Structures

Using a combination of microwave irradiation and ILs, other complex metal structures with enhanced performance were synthesized. Cybinska et al. prepared Ln^{3+}-doped (Ln = Sm, Eu, Tb, Dy) $BiPO_4$ NPs with particle sizes <10 nm from the respective metal acetates and choline or butylammonium dihydrogen-phosphate in a microwave at 120 °C for 10 min [76]. In their work, ILs acted as a solvent, microwave susceptor and reaction stabilizer; they also affected particle morphology and size. A big advantage of MAIL methods is that no post processing (including heat treatment) of the final product is needed, which helps to avoid particles aggregation.

newpage

Siramdas et al. successfully synthesized luminescent InP nanocrystals with different sizes using a 1-butyl-4-methylpyridinium tetrafluoroborate ([BmPy][BF_4]) IL [77]. The resulting products exhibited color-tunable luminescence in the visible region with 20%–30% quantum yields. They compared products obtained without an IL in the reaction system (see Figure 7) and concluded that IL presence was a key factor responsible for efficient nanocrystals luminescence. This simple procedure provides a pathway to obtain luminescent nanocrystals using a simple preparation method with just a single precursor.

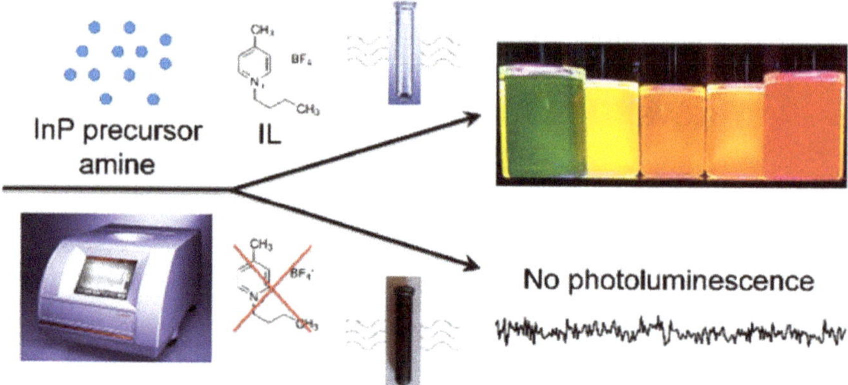

Figure 7. Scheme of microwave-assisted synthesis of InP nanocrystals in ionic liquid (IL) with color-tunable luminescence [77]. Copyright American Chemical Society, 2017.

Olchowka et al. fabricated fluoridosilicates A_2SiF_6 (A = Li, Na, K, Rb, Cs) by the MAIL method using 1- butyl-3-methylimidazolium hexafluorophosphate ([Bmim][PF_6]). These materials are promising host lattices for LEDs [78]. During their synthesis, an IL was used as a low-temperature solvent and as a fluorine source, which helped to eliminate usage of hazardous HF. They suggested expanding usage and variety of different ILs as well as reaction time to explore this green method to obtain a variety of fluoride-containing LED host materials with different particle sizes and shapes as well as functionalities.

Various ILs were used by Alammar et al. to synthesize nano-$SrSnO_3$ photocatalysts using microwaves [79]. They used ILs with different cations (e.g., 1-butyl-3-methylimidazolium ([C_4mim]$^+$), 6-bis(3-methylimidazolium-1-yl)hexane ([C_6(mim)$_2$]$^{2+}$), butylpyridinium ([C_4Py]$^+$), tetradecyltrihexylphosphonium ([P_{66614}]$^+$)) and bis(trifluoromethanesulfonyl)amide ([Tf_2N]$^-$) as an anion. They concluded that usage of different ILs could lead to products with different morphologies and sizes. They also emphasized, that unlike the IL cation, the choice of anion had little effect on morphologies of the formed products.

Besides the mentioned references above, Table 1 lists an updated review of other metal structures prepared by microwave-assisted ionic-liquid (MAIL) methods reported in the literature.

Table 1. An updated review of other metal structures prepared by microwave-assisted ionic-liquid (MAIL) methods reported in the literature.

Nanomaterial Properties.	ILs	Microwave Condition	Applications	Ref.
One-dimensional (1D) and 2D mesoporous nickel cobaltite ($NiCo_2O_4$) rods and sheets	[Bmim][BF_4]	100 °C for 10 min	Efficient electrode material for supercapacitors	[80]
Metal-fluoride NPs, (MFx-NPs) with M = Fe, Co, Pr, Eu	[Bmim][BF_4]	Irradiated for 10 min (Co) or 15 min (Fe, Pr, Eu) at a power of 50 W to a temperature of 220 °C	Cathode material for lithium-ion batteries	[81]
Nanoparticle morphology of Sb_2Te_3	1-alkyl-3-methylimidazolium and 1,3-dialkylimidazolium-based ILs	30 s at 100 °C, subsequently for 5 s at 150 °C and finally for 5 min at 170 °C	Thermoelectrics	[82]
Hierarchical microcube-like BiOBr	1-hexadecyl-3-methylimidazolium-bromide	160 °C	/	[83]
$Sr_{1-x}Ba_xSnO_3$ Perovskite	[C_4mim][Tf_2N]	10 min at 85 °C	Photocatalytic applications for the hydroxylation of terephthalic acid	[84]
Co_2P/CNTs	Tetrabutylphosphonium chloride ([$P_{4,4,4,4}$]Cl), trihexyl(tetradecyl)phosphonium chloride ([$P_{6,6,6,14}$]Cl)	Microwave oven for 6 min	Hydrogen evolution	[85]
$SnSe_x$ NDs@rGO	[Bmmim]Cl	120 °C for 5 min followed by 180 °C for 55 min	Electrochemical application	[86]
MoS_2/BiOBr	[C_{16}mim]Br	140 °C for 10 min	Photocatalyst	[87]

3. Polymers Preparation

In order to reduce consumption of volatile organic compounds (VOCs), the polymer industry is in search for alternatives, some of which are water or supercritical CO_2 [88]. However, some polymer-related processes cannot be exposed to moisture, and practices involving supercritical CO_2 require high pressure [89]. Another alternative is the implementation of ILs as polymer reaction media. Polymerization reactions conducted in ILs exhibited higher polymerization rate and higher molar weight of the resulting polymers. The application of microwave irradiation in the procedures involving polymers also attracts attention because of its higher efficacy compared to conventional heating methods. Thus, the combination of ILs and MW could promote development of greener and more efficient polymer reactions with significant energy savings as well as with reduced or completely eliminated VOCs [90].

3.1. Ring Opening Polymerization

Cationic ring opening polymerization (CROP) is an easy route to synthesize poly(2-oxazoline)s (POxs) [91–93]. POxs are functionalized polymers with various applications, however, they often have low polymerization rates and high VOC emissions during their production which limits their vast development [94–96]. The first study showing the application of ILs as reaction media for CROP under MW was reported by Guerrero-Sanchez et al. [97]. Even though they performed CROP of 2-ethyl-2-oxazoline in different ILs, their main focus was CROP in a [Bmim][PF_6] IL because poly(2-ethyl-2-oxazoline) is more soluble in water than an IL which can improve polymer isolation and IL recycling. The polymerization process was finished within 1 min under MW, and faster polymerization rate and lower polydispersity was observed during this reaction in comparison with reactions using other common organic solvents. Additionally, the [Bmim][PF_6] IL was efficiently recovered by water extraction and reused with similar results. They also successfully synthesized 2-phenyl-2-oxazoline and 2-(m-difluorophenyl)-2-oxazoline in water-soluble 1-butyl-3-methylimidazolium trifluoromethanesulfonate ([Bmim][TfO]) and [Bmim][BF_4] ILs in just 30 min by microwave-assisted reaction [98]. Their results showed that polymers obtained in ILs demonstrate broad molecular weight distributions and higher polydispersity.

Poly(ε-caprolactone) (PCL) is one of the most common industrial biodegradable materials for bio- and medical packaging applications. Ring opening polymerization (ROP) of ε-caprolactone(ε-CL) to obtain PCL requires over 24 h when conventional heating is used [99]. Liao et al. obtained PCL in just 30 min by performing ROP with ZnO as a catalyst and under the presence of [Bmim][BF_4]. They observed that at the same microwave power, temperature of the reaction mixtures can be increased by just varying concentration of [Bmim][BF_4]. Additionally, ROP of ε-CL was also successfully performed without a catalyst but with [Bmim][BF_4] at higher temperature. The same team studied microwave-assisted ROP of another biodegradable material—poly(trimethylene carbonate) (PTMC) [100]. When 5 wt.% of [Bmim][BF_4] was added, PTMC with molar weight equal to ~36,400 g/mol was obtained at 5 W microwave power in 60 min. Molar weight and reaction rate were both increased when the combination of ILs and MW were used.

3.2. Polycondensation

Polycondensation reactions using the MAIL technique were mainly reported by Mallakpour's team. They prepared chiral poly(amide–imide) nanostructures, active poly(ester–imide) and other polyamides [101,102] using tetrabutylammonium bromide (TBAB), 1,3-diisopropylimidazolium bromide([1,3-(isopr)$_2$im]Br) ILs, etc [103]. In a typical reaction, 5-(2-Phthalimidiylpropanoyl-amino) isophthalic acid, diamines [1,3-(isopr)$_2$im]Br and 1,5-Naphthalenediamine or other aromatic diamines were placed in a porcelain dish and then triphenyl phosphite (TPP) was added [104]. The mixtures were heated under microwave irradiation for a total of 90 s. After filtering and drying, the yield was 93%. They also compared their MAIL methods to conventional heating ones. Products obtained under microwave irradiation had remarkable solubility, optically activity and good thermal stability. At the same time, reaction time was reduced remarkably, and ILs used were harmless green solvents. Thus, the techniques developed by Mallakpour's team are environmentally friendly polycondensation reactions.

3.3. Free Radical Polymerization

During heterogeneous free-radical polymerization (FRP) reactions, ILs can also act as reaction media, as surfactants and as microwave absorbents. Guerrero-Sanchez et al. first investigated FRP of methyl methacrylate in [Bmim][TfO] and [Bmim][BF_4] ILs under microwave irradiation [98]. ILs in this reaction provided a more efficient absorption of microwave irradiation and increased the heating rates. At the same time, because of the dispersed phase, type of IL and its concentration also influenced yield and molecular weight of the resulting polymer. After the reaction, ILs were recovered and reused, which confirmed that MAIL-based FRP reactions are not only efficient but also environmentally friendly.

Glück's team studied systematic variation of MAIL FRP reaction conditions to compare them with conventional heating methods and solvents [105,106]. They found that when conventional solvents (such as N,N-Dimethylformamide(DMF) or methanol) were used, conversion rates of the FRP reaction conducted using microwaves and using conventional heating were the same. However, when ILs were used as solvents, FRP conversion rate upon microwave heating was higher relative to conventional heating whether with or without ILs. Polymerization rate r_p was higher in ILs than in conventional solvents upon microwave irradiation but only for some reactions. In other reactions, r_p was nearly the same in ILs and in DMF/methanol. These differences might be because of complexity of polar interactions between ILs and the monomer and/or polymer radicals.

3.4. Copolymers

3.4.1. Cellulose Graft Copolymers

Intelligent polymers with unique properties in response to external triggers are attractive for biotechnology. Grafting such intelligent polymers on cellulose was extensively studied. Using dissolving capacity of cellulose in [Bmim]Cl, Wei et al. grafted 2-(Dimethylamino) ethylmethacrylate (DMAEMA) onto bagasse cellulose by microwave heating [107]. This cellulose-g-DEAEMA copolymer was sensitive to pH and salinity in an acidic environment or a low salt concentration. A homogeneous cellulose–acrylic acid graft copolymer as an effective metal ion adsorbent was prepared by Lin et al. [108] using [Bmim]Cl as a powerful direct solvent. Usage of microwave irradiation reduced the time of this reaction to only 3 min. After optimization of reaction parameters (such as monomer concentration, exposure temperature and overall reaction time), a graft copolymer with excellent adsorption ability towards Cu^{2+} (to remove it from industrial waste and contaminated waters) was obtained.

3.4.2. Copolyurethanes

Segmented polyurethanes (PUs), used in a variety of practical applications, in which structure variability is needed, are copolymers with hard and soft segments. Rafiemanzelat et al. synthesized novel multi-block polyurethane copolymers, containing ecofriendly and biodegradable segments, by two-step polymerization in TBAB under microwave irradiation [109]. First, 4,4′-methylene-bis-(4-phenylisocyanate) was reacted with L-leucine anhydride cyclodipeptide to produce isocyanate-terminated oligo(imide–urea) as a hard segment. Then, polyethylene glycol with a molecular weight of ~1000 was used to extend the chain of the pre-polymer for further synthesis of PU. Comparing to conventional heating methods, PUs prepared by the MAIL technique showed more phase separation and crystallinity. Furthermore, reusability of ILs was also confirmed in their work.

3.5. Poly(ionic Liquid)s

Poly(ionic liquid)s (PILs) can be synthesized by polymerization of IL monomers. PILs have a lot of unique properties: They inherit some properties of ILs (such as low vapor pressure and high thermostability) and possess well-defined solid morphologies and mechanical characteristics [110] of the original polymers. Although conductivity of PILs decreases after polymerization, PILs are still suitable for electro-responsive fluid applications because of locally-moving ions and the high density of cation/anion parts [111–113]. Dong et al. prepared uniform P[MTMA][TFSI] (MTMA = [2-(methacryloyloxy)ethyl]trimethylammonium$^+$, TFSI = bis(tri-uoromethane sulfonyl)imide$^-$) particles with low density under microwave irradiation for 7 h at 70 °C (see Figure 8) [112]. Their PILs showed high electro-responsive fluid activity without the need for an activator and low leaking current density because of their hydrophobicity. Analysis of dielectric spectra showed that high density of cation/anion parts in PILs played an important role in strong dielectric polarizability and adequate polarization response of PILs. Overall, PIL particles have a lot of potential applications in electro-responsive systems. Based on previous research [105], Dong et al. further demonstrated electrorheological applications of PIL particles as well as studying how sizes of cation/anion parts influence electrorheological performance [113]. Monodisperse PIL particles with a polystyrene backbone and quaternary ammonium/fluorinated imides as ion parts were synthesized under microwave irradiation. Electrorheological effect was stronger when cation/anion parts were smaller, probably because of the higher density of ions and faster ion mobility rates.

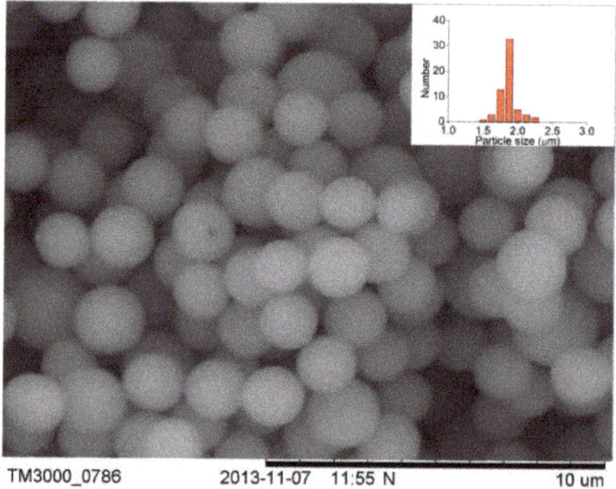

Figure 8. SEM image and size distribution (inset) of P[2-(methacryloyloxy)ethyl]trimethylammonium$^+$ bis(tri-uoromethane sulfonyl)imide$^-$) (P[MTMA][TFSI]) particles [112]. Copyright Royal Society of Chemistry, 2014.

Based on the hydrophilicity and good extraction efficiency of some PILs, Yang's team designed PIL-modified magnetic nanomaterials under microwave assistance for pesticide and phthalate esters extraction [114,115]. PIL (1-vinyl-3-butylimidazolium bromide)-functionalized modified polystyrene magnetic nanospheres (PILs-PMNP) were synthesized by a one-pot copolymerization reaction under microwave irradiation for 4 h. PILs-PMNPs were designed for magnetic solid-phase extraction (SPE) to extract and concentrate phthalate esters from beverages. These PILs-PMNPs demonstrated outstanding adsorption ability toward traces of PAE analytes compared with polystyrene-modified magnetic nanospheres. Yang et al. also designed PIL-modified magnetic nanoparticles based on a dispersion SPE method, which was combined with HPLC–UV to detect trace levels of pesticides in fruits and vegetables [115]. PIL materials could also be reused multiple times.

4. Preparation of Carbon-Based Composites

Carbon-based composites are widely used as catalysts, electrode and luminescent materials as well as super capacitors [21]. Carbon-based composites are typically synthesized by pyrolysis of raw biomaterials or polymers at high temperatures [116,117]. IL-based approaches to prepare carbon-based composites also attracted attention [118] mostly because of the unique properties of ILs [119,120]. For instance, ILs could be used as solvents for the synthesis of carbon materials from biomass [121], or IL cations or anions could act as porogens to create porous carbons [122], or even as carbon or nitrogen sources to produce carbon-based composites [123]. All these and similar methods when combined with microwave irradiation offer even more effective and economic methods to develop and design various carbon-based composites.

4.1. Carbon Dots/Nanodots

Carbon dots (CDs) have excellent luminescent properties, good solubility and biocompatibility because of their quantum properties [124]. Conventional CD syntheses usually occur at high temperatures. Usage of ILs could offer a larger temperature range for CD synthesis because of the unique physical and chemical properties of ILs [125]. ILs can also be used as solvents, carbon sources, modifiers for CD syntheses, etc. Microwave irradiation, which could be combined with ILs, can also be implemented to synthesize CDs especially with biomaterials as initial compounds [123]. Jeong et al. reported a novel approach to produce CDs from biomaterials [126] using just microwave irradiation: They fabricated CDs based on regenerated cellulose (RC–CDs) from recycled paper. They combined microwave irradiation with a 1-allyl-3-methylimidazolium chloride ([Amim]Cl) IL to extract cellulose from the recycled paper, which was then dissolved in the IL. Cellulose was then regenerated by adding absolute ethanol, after which the IL was recovered. RC–CDs exhibited low cytotoxicity and excellent fluorescence properties for potential bio-applications.

Rice straw, a common bio-resource, was also used as a raw material to fabricate photo-luminescing CDs by the MAIL method with an [Amim]Cl IL [127]. The IL helped to extract and dissolve cellulose from the rice straw and provided nitrogen as a heteroatom to obtain N-doped CDs. The resulting CDs were spherical and had a high quantum yield. Moreover, they were also used as efficient and very responsive Fe^{3+} sensors with a very low detection limit.

Very recently, Pham-Truong et al. used glutamine and glucose as carbon and nitrogen sources, respectively, to synthesize nano-CDs by the MAIL method using a 1-ethyl-3-methylimidazolium ethylsulfate ([Emim][$EtSO_4$]) IL [125]. The resulting products demonstrated presence of pyridinic and graphitic nitrogen as well as imidazolium and the IL anion on their surfaces. The source of graphitic nitrogen was from the imidazolium ring in the IL. Oxygen reduction measurements showed that both CDs have promising electrocatalytic activity. However, CDs generated under the presence of glucose and the IL exhibited the highest efficiency (up to 90%) towards H_2O_2 production.

Liu et al. also used ILs as a carbon source to prepare fluorescence CDs by one-step microwave-assisted pyrolysis of a mixture containing a 1-vinyl-3-aminopropyl imidazolium ([PAVIm]Br) IL and ethylenediamine [128]. During their synthesis, the IL acted as a carbon source and ethylenediamine served as a nitrogen source for doping and as a surface passivating agent. The resulting CDs exhibited outstanding photoluminescence properties useful for detection of Cr(VI) as well as for temperature and pH values. Safavi et al. suggested adopting the MAIL method for a simple and efficient preparation of blue-luminescing CDs with narrow particle size distribution [129]. They used a N-octylpyridinum hexafluorophosphate IL as a microwave absorber and as a carbon source. The same CDs were used for effective catalytic degradation of azo dyes.

4.2. Carbon Nanoparticles

Carbon nanoparticles (CNPs) are other fluorescent materials used in many applications because of their outstanding properties such as fluorescence, low cytotoxicity and excellent biocompatibility.

Xiao et al. group was the first to report microwave-assisted one-step synthesis of CNPs derived from ILs. Their CNPs were monodisperse with strong fluorescence and good water solubility [130]. They employed a [Bmim][BF_4] IL as a carbon and nitrogen source as well as different microwave irradiation times to design various CNPs. The whole process was eco-friendly and economical without any complex procedures or toxic reactants. These as-prepared CNPs were incorporate into a novel fluorescence probe for quercetin detection. Wei et al. fabricated a CNP/ionic liquid (CNPIL) hybrid by exposing 1-butyl-3-methylimidazolium glutamine salt (as the IL) and glucose to microwave irradiation [131]. The resulting CNPIL showed excellent conductivity and dispersibility in organic and/or aqueous solvents. These CNPIL also demonstrated outstanding photoluminescence properties adjustable by specific synthesis conditions.

4.3. Other Carbon Materials

Meng et al. used IL 1-butyl-3-methylimidazolium dicyanamide ([Bmim][$N(CN)_2$]) as a carbon source to prepare a N-doped carbon membrane under microwave irradiation. This membrane had high discharge capacity and excellent cyclic performance especially relative to membranes synthesized from glucose as a carbon source [132]. Because of the absorption properties of the IL, a uniform and network-like N-doped carbon membrane was coated on the surface of $LiFePO_4$ to improve its electronic conductivity. Vadahanambi et al. reported a one-pot synthesis of holey-carbon nano-sheets (hCNS) by microwave carbonization of 1-ethyl-3-methylimidazolium tetrafluoroborate ([Emim][BF_4]) in a soft-template [133]. The resulting hCNS were homogeneously distributed spheres 10–20 nm in diameter, which could be used as potential anode materials for sodium ion batteries.

MAIL technology can also be applied to modify carbon materials. Beggs et al. improved interfacial adhesion of carbon fibers by treating them in the (1-ethyl-3-methylimidazolium bis(trifluoromethylsulfonyl)imide) IL under lower-power microwave irradiation (20 W) [134]. Myriam et al. reported modification of CNTs by a newly-developed approach for controlled functionalization/retro-functionalization of single-walled CNTs using the MAIL method. This method can be used as a general technique to restore π-conjugated structure of carbon nanotubes and other carbon allotropes [135].

5. Preparation of Biomass-Based Materials

It is widely known that cellulose dissolves in ILs, and microwave heating can significantly accelerate the dissolution. Synthesis of biomass-based nanocomposites and products using ILs significantly broadens biomass application range. Additionally, using biomass as a raw material satisfies green chemistry principles of sustainable resources.

5.1. Cellulose-Based Composites

Recently, cellulose-based composites attracted a lot of attention as green materials. These composites enrich the family of green materials and expand usage of biomass for high value-added applications. Combination of cellulose with metals yields even more outstanding properties useful for biomedical applications [136]. During the synthesis of cellulose-based composites, ILs not only act as templates and solvents, they also have outstanding contribution to the dissolving capacity of cellulose. Based on these properties Jia's group first developed MAIL technology to synthesize cellulose-based nanocomposites [137]. They prepared cellulose/calcium silicate nanocomposites using fresh and recycled ILs. They used [C_4mim]Cl as a solvent to dissolve microcrystalline cellulose and microwaved the solution to 110 °C for 20 min. $Ca(NO_3)_2{\cdot}4H_2O$ and $Na_2SiO_3{\cdot}9H_2O$ were added as precursors. SEM of the nanocomposites synthesized by fresh, recycled and twice-recycled ILs are shown in Figure 9. Sizes and microstructures of the final products were different depending on which IL was used. Their team also synthesized cellulose/CuO and cellulose/$CaCO_3$ using MAIL methods [138,139] with [Bmim][BF_4] and [Bmim]Cl ILs, respectively. Both of these ILs are excellent for dissolution of cellulose and as microwave absorbents. Microwave-heating time and IL:cellulose ratio significantly affected the size and morphologies of the nanocomposite. Cellulose/$CaCO_3$ nanocomposites synthesized using this method demonstrated good biocompatibility, which is useful for biomedical applications. Thus, this synthetic method opens a new way for cellulose applications.

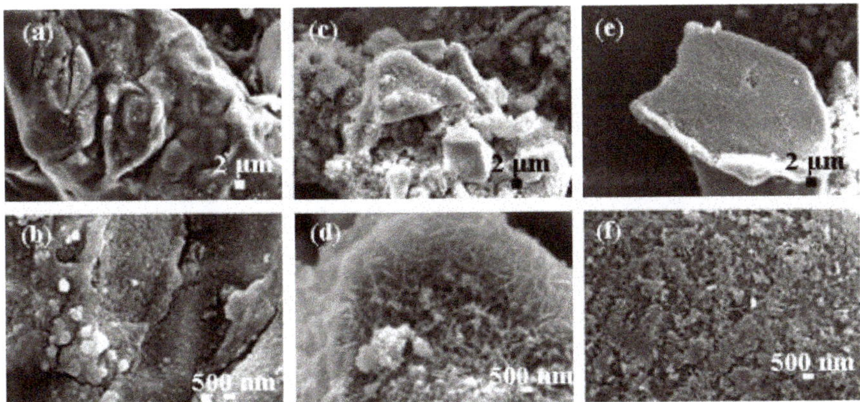

Figure 9. SEM micrographs of the cellulose/calcium silicate nanocomposites synthesized in (**a**,**b**) the starting ionic liquids, (**c**,**d**) the once-recycled ionic liquids, and (**e**,**f**) the twice recycled ionic liquids [137]. Copyright Elsevier, 2011.

Sharma et al. [140] prepared superabsorbent nanocomposite from sugarcane bagasse (SCB), chitin and clay by the MAIL method with a 2-hydroxy ethyl ammonium formate (HEAF) IL. SCB and chitin were first suspended in separate HEAF solutions and then allowed to precipitate. After mixing appropriate amounts of initiator, monomer and crosslinker, they obtained a nanocomposite with over 1000% swelling degree upon microwave irradiation for 24 h. They also found that these nanocomposites exhibited high microbial resistance and recyclability with a swelling degree even greater after each recycle.

Wood fibers can also be used as raw material to fabricate transparent film composites. Lu et al. obtained a highly transparent all-cellulose film with less time and better performance (than conventional techniques) by using MAIL technology [141]. They used an IL as a solvent to disrupt stiff structure of the wood fibers: the IL broke hydrogen bonds between or/and inner cellulose chains. Meanwhile, microwave irradiation increased the temperature of the system very quickly during the synthesis, which resulted in higher efficiency.

Cellulose-ZnO nanocomposite [142], cellulose-SnS_2 composites [143], cellulose/CaF_2 nanocomposites and cellulose/MgF_2 nanocomposites [144] were also successfully synthesized using the MAIL method. Thus, choice of IL and microwave conditions affect morphologies and properties of the resulting nanocomposites. This method of preparation of novel materials opens a lot of possibilities for current and future cellulose-based nanocomposites.

5.2. Production of 5-Hydroxymethylfurfural

An IL was also used in 5-hydroxymethylfurfural (5-HMF) production from biomass with high efficacy. 5-hydroxymethylfurfural (5-HMF) is one of the top bio-based platform compounds, which acts as an important intermediate product to obtain fine chemicals, pharmaceuticals and biofuels [145]. 5-HMF is also a key building block and a starting point for synthesis of 2,5-dimethylfuran, 2,5-diformylfuran, 2,5-furandicarbaldehyde and 2,5-furandicarboxylic acids (see Figure 10) [146]. 5-HMF can be obtained from fructose, glucose or cellulose by hydrolysis, dissolution and acid-assisted catalysis. However, the yield of 5-HMF when produced from biomass or cellulose is unsatisfactory.

Thus, ILs can be used in 5-HMF production from biomass because of its attractive cellulose dissolving ability. In fact, MAIL methods for 5-HMF production from cellulose were indeed reported [147].

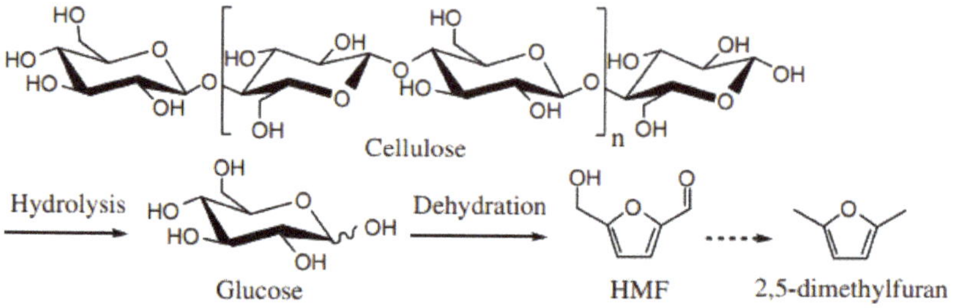

Figure 10. Schematic illustration of the steps for cellulose to HMF [148]. Copyright Elsevier, 2009.

Early reports on the application of MAIL method for 5-HMF production were generated by Prof. Li's and Qi's team. They both used $CrCl_3$ as a catalyst, 400 W MW as a heating source and $[C_4mim]Cl$ and [Bmim]Cl as ILs, respectively [148,149]. Their groups achieved 61% and 71% yields for 5-HMF, respectively. They used ILs as solvents for the acid-catalyzed dehydration. At the same time, chloride anions in ILs contributed to the isomerization of glucose into fructose. $[C_4mim]Cl$ also acted as a scavenger for water: Water is often responsible for low yields as it decomposes freshly-formed HMF during its synthesis. Compared to conventional heating methods, MW reduced reaction time to 30 s, which was possible because of excellent dielectric properties of ILs. Prof. Qi et al. [149] tested recyclability of the system and demonstrated that ILs/$CrCl_3$ could be recycled without compromising its performance.

Usage of ILs as solvents implies that high amounts IL are often needed during the synthesis. Qu et al. proposed to use ILs as catalysts (instead of as solvents) to convert microcrystalline cellulose to 5-HMF to make the process more economical and environmentally friendly [150]. The highest yield of 5-HMF was 28.63% when 1,1,3,3-tetramethylguanidine tetrafluoroborate ($[TMG][BF_4]$) was applied as a catalyst under microwave irradiation at 132 °C for 48 min. Very recently, Zhang et al. developed a process to produce 5-HMF generation from cellulose at mild temperatures (~80 °C) using microwave irradiation for 3 h [151]. Their catalyst was a fluorine anion-containing IL functionalized with biochar sulfonic acids (BCSA-IL-F_{1-3s}), which can be simply synthesized by ionic exchange of 1-trimethoxysilylpropyl-3-methylimidazolium chloride (IL-Cl) grafted on the BCSA with CF_3SO_3H (HF_1), HBF_4 (HF_2), HPF_6 (HF_3), respectively. The highest yield of 5-HMF was 12.70%–27.94%.

5.3. Production of Furfural

Furfural is an important high value-added platform chemical derived from lignocellulosic biomass, such as corncobs, bagasse, wood chips and switchgrass. Furfural has great platform potential for the simultaneous production of fuels including methyl-tetrahydrofuran and liquid alkanes [152]. Serrano-Ruiz et al. proposed an efficient pathway for the production of furfural from C5 sugars and lignocellulosic waste catalyzed by Brönsted acidic ionic liquids (1-(4-Sulfonylbutyl) pyridinium methanesulfonate) under microwave irradiation (100 W, 150 °C). The yield of furfural was 85% and 45% for C5 sugars and lignocellulosic waste, respectively [153]. After that, Zhang et al. studied the conversion of xylan and lignocellulosic biomass (corncob, grass and pine wood) into furfural catalyzed by $AlCl_3$ in [Bmim]Cl at 170 °C for 10 s microwave. The highest furfural production was 84.8% for xylan and 16%–33% for lignocellulosic biomass. Meanwhile, [Bmim]Cl and $AlCl_3$ could be recycled for four runs with stable catalytic activity [154].

Zhang's work focused on the conversion of lignocellulosic biomass into HMF and furfural at the same time in $[C_4mim]Cl$ in the presence of $CrCl_3$ under microwave irradiation [155]. Corn stalk, rice straw and pine wood treated under typical reaction conditions produced HMF and furfural in yields of 45%–52% and 23%–31%, respectively, within 3 min. da Silva Lacerda et al. also produced HMF and furfural from lignocellulosic materials with a yield of 53.24% at optimized condition [156]. The analysis result showed that when an additional 15 min ultrasonic pretreatment is also conducted, microwave treatment combined with stirring could result in the best production yields. The reports described above should be valuable to facilitate cost-effective conversion of biomass into biofuels.

5.4. Production of Reducing Sugar

Bio-ethanol energy, as a renewable resource, gets increasing attention recent years. Reducing sugar hydrolyzed from cellulose and hemicelluloses can be relatively easy fermented to bio-ethanol. The application of ILs and microwaves in biomass hydrolysis cannot only obtain a high yield of reducing sugar, but also achieves high efficiency. Chen et al. proposed an effective and profitable process with [Amim]Cl as catalyst to produce reducing sugars from the hydrolysis of straw under microwave heating [157]. The highest yield of glucose and xylose can attain 29.1% and 24.3%, respectively. Hou et al. also achieved high sugar yield by optimized microwave-assisted [TBA][OH] pretreatment [158]. They suggested that the combined effect leads to the violent deconstruction of lignin and hemicelluloses, the crystalline region broken and an eroded, pored and irregular micro-morphology, which contribute to the high sugar yield.

6. Mechanisms Involved in MAIL Methods

There are several mechanisms involved in MAIL methods for materials preparation. First of all, ILs efficiently absorb microwave energy through an ionic conduction mechanism, and thus are employed as solvents and co-solvents, leading to a very high heating rate and a significantly shortened reaction time. Yin's group also revealed that even a small amount of IL could act as microwave absorber to markedly shorten the reaction time [73]. The above-mentioned results all demonstrate that the reactions could be completed in a few minutes by means of MAIL methods.

Second, ILs are always used as solvents in biomass-based materials and carbon materials syntheses mainly due to the cellulose dissolving capacity of ILs. For instance, [Bmim]Cl is usually applied in biomass dissolution. The imidazolium cation with positive charge acting as electron acceptor and the chloride anion with negative charge acting as electron donor, interact with oxygen and hydrogen of -OH bonding of cellulose, respectively and promote the dissolution of cellulose. The hydroxyl groups on the cation side chain of the ionic liquid along with the anion also interact with the hydroxyl groups of cellulose, which further weakens hydrogen bonding among the cellulose fibers. Ionic liquids weaken the hydrogen bonding in cellulose which results in the formation of amorphous regions with a lower degree of order in the cellulose structure. Dissolution of the amorphous region takes place initially, followed by reaction of ionic liquid with the crystalline part of cellulose [159]. Based on this property, various materials were obtained.

Third, ILs are often used as additives in metal structures preparation due to their templating effect. There are many examples that MAIL method could resulted in different morphologies of the obtained nanostructured materials above. Ionic liquids play a critical role in the formation of various structures. Among the reports, imidazolium rings caused great attention in such reactions. The extended hydrogen-bonding and π-π stack interaction of the neighboring imidazolium rings enable ionic liquid molecular recognition and self-assembly. As a "supramolecular" solvent, the self-assembled ability of the IL significantly influences structural orientation in the reaction [160]. Pang et al. proposed the formation of different morphologies from another perspective. These ionic liquids have a cation head and carbon chain group, therefore they have the aggregation behavior similar to the surfactant. The aggregates in the solution might also have a function like the soft template for the growth of the inorganic crystals [161]. Microwave irradiation in some reports accelerates the migration rate of atoms which contribute to different structures compared with conventional heating method.

Last, parts of ILs also provide carbon, nitrogen and fluorine resources to the obtained materials, such as ILs could offer a carbon source to prepare fluorescence CDs [128] and in Chen's study, the IL provided a fluorine-rich environment and helped to expose (001) facets of TiO_2 [60].

7. Conclusions

The combination of microwave irradiation and ionic liquids (ILs) paves a powerful route towards high efficiency and low toxicity processes. In this review, applications of ILs in conjunction with microwave irradiation to prepare different inorganic, polymeric, carbon- and biomass-based materials were highlighted. ILs can be used as microwave absorbers, solvents, additives, templates and reactants. As ionic conductors, ILs absorb microwaves, which benefits heating processes. As solvents, ILs are low toxicity and, at the same time, are outstanding for cellulose dissolution. When used as additives, ILs can broaden molecular weight distribution of the resulting polymers. As templates, different types of ILs with different doses could control morphologies and properties of the resulting nanostructures. As reactants, ILs act as abundant carbon and nitrogen sources. Furthermore, some researchers suggested that ILs are recyclable and can be reused without loss of their performance.

Despite numerous achievements, several limitations still exist in the development of MAIL technology. First, only limited numbers of ILs were applied to prepare common materials, such as [Bmim]Cl and [Bmim][BF_4]. Development and application of other none toxic types of ILs still need further scientific exploration. Second, there is still a lack of knowledge on the exact mechanism between microwaves, ILs and nanostructures or polymers. Thus, future research should focus on providing more experimental examples for better understanding these interactions. Third, recyclability of ILs has very important economic as well as environmental benefits for current and future technologies. Yet, it is still challenging to comprehend all the details of the processes and reactions related to ionic liquid recovery process. Finally, it is also essential to develop the large-scale industrial application of this method in the future.

In summary, this review provided literature-supported evidence of advantages of synthesis techniques involving microwave irradiation of IL-containing systems to obtain various functional materials. The goal of this review paper was not only to convey remarkable recent developments of this synthetic technology, but also to inspire the readers to study this novel and effective route for the new materials preparation even further.

Author Contributions: Y.W. wrote the original draft and did the literature research for the paper; Q.H. gave some guidance in writing a review; M.J. provided some articles and studying materials; W.L. contributed literature study and supervised the work.

Funding: This research was supported by the National Natural Science Foundation of China (21878163), the National Key Research Project (2018YFD0800103), the Fundamental Research Funds for the Central Universities, and the Natural Science Foundation of Tianjin, China (17JCZDJC39500).

Conflicts of Interest: The authors declare no conflict of interest.

References

1. Rogers, R.D. CHEMISTRY: Ionic liquids—Solvents of the future? *Science* **2003**, *302*, 792–793. [CrossRef] [PubMed]
2. Torimoto, T.; Tsuda, T.; Okazaki, K.; Kuwabata, S. New frontiers in materials science opened by ionic liquids. *Adv. Mater.* **2010**, *22*, 1196–1221. [CrossRef]
3. Hapiot, P.; Lagrost, C. Electrochemical reactivity in room-temperature ionic liquids. *Chem. Rev.* **2008**, *108*, 2238–2264. [CrossRef]
4. Dupont, J. From molten salts to ionic liquids: A "nano" journey. *Acc. Chem. Res.* **2011**, *44*, 1223–1231. [CrossRef]
5. Liu, H.; Liu, Y.; Jinghong, L. Ionic liquids in surface electrochemistry. *Phys. Chem. Chem. Phys.* **2010**, *12*, 1685–1697. [CrossRef]

6. Parnham, E.R.; Morris, R.E. Ionothermal synthesis of zeolites, metal-organic frameworks, and inorganic-organic hybrids. *Acc. Chem. Res.* **2007**, *40*, 1005–1013. [CrossRef]
7. Muldoon, M.J.; Nockemann, P.; Lagunas, M.C. Crystal engineering with ionic liquids. *CrystEngComm* **2012**, *14*, 4873. [CrossRef]
8. Zhao, Q.; Chu, H.; Zhao, B.; Liang, Z.; Zhang, L.; Zhang, Y. Advances of ionic liquids-based methods for protein analysis. *TrAC Trends Anal. Chem.* **2018**, *108*, 239–246. [CrossRef]
9. Morris, R.E. Ionothermal synthesis—Ionic liquids as functional solvents in the preparation of crystalline materials. *Chem. Commun.* **2009**, *0*, 2990–2998. [CrossRef]
10. Kaur, N.; Singh, V. Current status and future challenges in ionic liquids, functionalized ionic liquids and deep eutectic solvent-mediated synthesis of nanostructured TiO_2: A review. *New J. Chem.* **2017**, *41*, 2844–2868. [CrossRef]
11. Cevasco, G.; Chiappe, C. Are ionic liquids a proper solution to current environmental challenges? *Green Chem.* **2014**, *16*, 2375–2385. [CrossRef]
12. Dai, Y.; van Spronsen, J.; Witkamp, G.; Verpoorte, R.; Choi, Y.H. Natural deep eutectic solvents as new potential media for green technology. *Anal. Chim. Acta* **2013**, *766*, 61–68. [CrossRef]
13. Chen, Y.; Zhang, X.; You, T.; Xu, F. Deep eutectic solvents (DESs) for cellulose dissolution: A mini-review. *Cellulose* **2019**, *26*, 205–213. [CrossRef]
14. Gutowski, K.E.; Holbrey, J.D.; Rogers, R.D.; Dixon, D.A. Prediction of the formation and stabilities of energetic salts and ionic liquids based on ab initio electronic structure calculations. *J. Phys. Chem. B* **2005**, *109*, 49, 23196–23208. [CrossRef]
15. Namboodiri, V.V.; Varma, R.S. Solvent-free sonochemical preparation of ionic liquids. *Org. Lett.* **2002**, *4*, 3161–3163. [CrossRef]
16. Vidal, L.; Riekkola, M.L.; Canals, A. Ionic liquid-modified materials for solid-phase extraction and separation: A review. *Anal. Chim. Acta* **2012**, *715*, 19–41. [CrossRef] [PubMed]
17. Wang, H.; Wang, J.; Zhang, S.; Xuan, X. Structural effects of anions and cations on the aggregation behavior of ionic liquids in aqueous solutions. *J. Phys. Chem. B* **2008**, *112*, 16682–16689. [CrossRef] [PubMed]
18. Park, H.; Lee, Y.; Choi, B.; Choi, Y.; Yang, J.; Hong, W. Green one-pot assembly of iron-based nanomaterials for the rational design of structure. *Chem. Commun.* **2009**, *27*, 4058–4060. [CrossRef]
19. Li, M.; Dong, W.; Liu, C.; Liu, Z.; Lin, F. Ionic liquid-assisted synthesis of copper oxalate nanowires and their conversion to copper oxide nanowires. *J. Cryst. Growth* **2008**, *310*, 4628–4634. [CrossRef]
20. Swadźba-Kwaśny, M.; Chancelier, L.; Ng, S.; Manyar, H.G.; Hardacre, C.; Nockemann, P. Facile in situ synthesis of nanofluids based on ionic liquids and copper oxide clusters and nanoparticles. *Dalton Trans.* **2012**, *41*, 219–227. [CrossRef] [PubMed]
21. Xie, Z.; Su, D. Ionic liquid based approaches to carbon materials synthesis. *Eur. J. Inorg. Chem.* **2015**, *7*, 1137–1147. [CrossRef]
22. Paraknowitsch, J.P.; Zhang, J.; Su, D.; Thomas, A.; Antonietti, M. Ionic liquids as precursors for nitrogen-doped graphitic carbon. *Adv. Mater.* **2010**, *22*, 87–92. [CrossRef] [PubMed]
23. Meng, L.; Wang, B.; Ma, M.; Lin, K. The progress of microwave-assisted hydrothermal method in the synthesis of functional nanomaterials. *Mater. Today Chem.* **2016**, *1*, 63–83. [CrossRef]
24. Lidström, P.; Tierney, J.; Wathey, B.; Westman, J. Microwave assisted organic synthesis. *Tetrahedron* **2005**, *57*, 9225–9283. [CrossRef]
25. Wiesbrock, F.; Hoogenboom, R.; Schubert, U.S. Microwave-assisted polymer synthesis: State-of-the-art and future perspectives. *Macromol. Rapid Commun.* **2010**, *25*, 1739–1764. [CrossRef]
26. Hajipour, A.R.; Rafiee, F.; Ruoho, A.E. A rapid and convenient method for the synthesis of aldoximes under microwave irradiation using in situ generated ionic liquids. *J. Iran. Chem. Soc.* **2010**, *7*, 114–118. [CrossRef]
27. Zhang, C.; Liao, L.; Gong, S. Recent developments in microwave-assisted polymerization with a focus on ring-opening polymerization. *Green Chem.* **2007**, *9*, 303–314. [CrossRef]
28. Li, L.; Zhu, Y.; Ma, M. Microwave-assisted preparation of calcium sulfate nanowires. *Mater. Lett.* **2008**, *62*, 4552–4554. [CrossRef]
29. Zhu, Y.; Chen, F. Microwave-assisted preparation of inorganic nanostructures in liquid phase. *Chem. Rev.* **2014**, *114*, 6462–6555. [CrossRef]
30. Baig, R.B.N.; Varma, R.S. Alternative energy input: Mechanochemical, microwave and ultrasound-assisted organic synthesis. *Chem. Soc. Rev.* **2012**, *41*, 1559–1584. [CrossRef]

31. Polshettiwar, V.; Nadagouda, M.N.; Varma, R.S. The synthesis and applications of a micro-pine-structured nanocatalyst. *Chem. Commun.* **2008**, *47*, 6318–6320. [CrossRef] [PubMed]
32. Polshettiwar, V.; Baruwati, B.; Varma, R.S. Self-assembly of metal oxides into three-dimensional nanostructures: Synthesis and application in catalysis. *ACS Nano* **2009**, *3*, 728–736. [CrossRef]
33. Butt, F.K.; Bandarenka, A.S. Microwave-assisted synthesis of functional electrode materials for energy applications. *J. Solid State Electrochem.* **2016**, *20*, 2915–2928. [CrossRef]
34. Manzoli, M.; Gaudino, C.E.; Cravotto, G.; Tabasso, S.; Baig, R.B.N.; Colacino, E.; Varma, R.S. Microwave-assisted reductive amination with aqueous ammonia: Sustainable pathway using recyclable magnetic nickel-based nano-catalyst. *ACS Sustain. Chem. Eng.* **2019**, *7*, 5936–5974. [CrossRef]
35. Kappe, C.O. Controlled microwave heating in modern organic synthesis. *Angew. Chem. Int. Ed.* **2004**, *43*, 6250–6284. [CrossRef]
36. Varma, R.S.; Namboodiri, V.V. An expeditious solvent-free route to ionic liquids using microwaves. *Chem. Commun.* **2001**, *7*, 643–644. [CrossRef]
37. Lévêque, J.M.; Cravotto, G. Microwave, power ultrasound and ionic liquids: A new synergy in green organic synthesis. *CHIMIA Int. J. Chem.* **2010**, *60*, 613–620. [CrossRef]
38. Duan, X.; Ma, J.; Lian, J.; Zheng, W. The art of using ionic liquids in the synthesis of inorganic nanomaterials. *CrystEngComm* **2014**, *16*, 2550–2559. [CrossRef]
39. Ahmed, E.; Breternitz, J.; Groh, M.F.; Ruck, M. Ionic liquids as crystallisation media for inorganic materials. *CrystEngComm* **2012**, *14*, 4874–4885. [CrossRef]
40. Zhu, Y.; Wang, W.; Qi, R.; Xi, H. Microwave-assisted synthesis of single-crystalline tellurium nanorods and nanowires in ionic liquids. *Angew. Chem. Int. Ed. Engl.* **2004**, *43*, 1410–1414. [CrossRef]
41. Richter, K.; Campbell, P.S.; Baecker, T.; Schimitzek, A.; Yaprak, D.; Mudring, A.V. Ionic liquids for the synthesis of metal nanoparticles. *Phys. Status Solidi B* **2013**, *250*, 1152–1164. [CrossRef]
42. Li, Z.; Liu, Z.; Zhang, J.; Han, B.; Du, J.; Gao, Y.; Jiang, T. Synthesis of single-crystal gold nanosheets of large size in ionic liquids. *J. Phys. Chem. B* **2005**, *109*, 14445–14448. [CrossRef]
43. Bhawawet, N.; Essner, J.B.; Atwood, J.L.; Baker, G.A. On the non-innocence of the imidazolium cation in a rapid microwave synthesis of oleylamine-capped gold nanoparticles in an ionic liquid. *Chem. Commun.* **2018**, *54*, 7523–7526. [CrossRef]
44. Li, X.; Liu, M.; Cheng, H.; Wang, Q.; Miao, C.; Ju, S.; Liu, F. Development of ionic liquid assisted-synthesized nano-silver combined with vascular endothelial growth factor as wound healing in the care of femoral fracture in the children after surgery. *J Photochem. Photobiol. B* **2018**, *24*, 385–390. [CrossRef] [PubMed]
45. Safavi, A.; Tohidi, M. Microwave-assisted synthesis of gold, silver, platinum and palladium nanostructures and their use in electrocatalytic applications. *J. Nanosci. Nano.* **2014**, *14*, 7189–7198. [CrossRef]
46. Schütte, K.; Meyer, H.; Gemel, C.; Barthel, J.; Fischer, R.; Janiak, C. Synthesis of Cu, Zn and Cu/Zn brass alloy nanoparticles from metal amidinate precursors in ionic liquids or propylene carbonate with relevance to methanol synthesis. *Nanoscale* **2014**, *6*, 3116–3126. [CrossRef]
47. Schütte, K.; Barthel, J.; Endres, M.; Siebels, M.; Smarsly, B.M.; Yue, J.; Janiak, C. Synthesis of metal nanoparticles and metal fluoride nanoparticles from metal amidinate precursors in 1-butyl-3-methylimidazolium ionic liquids and propylene carbonate. *ChemistryOpen* **2017**, *6*, 137–148. [CrossRef]
48. Choi, S.M.; Seo, M.H.; Kim, H.J.; Kim, W.B. Synthesis and characterization of graphene-supported metal nanoparticles by impregnation method with heat treatment in H_2 atmosphere. *Synth. Met.* **2011**, *161*, 2405–2411. [CrossRef]
49. Lee, J.Y.; Yung, T.Y.; Liu, L.K. The microwave-assisted ionic liquid nanocomposite synthesis: Platinum nanoparticles on graphene and the application on hydrogenation of styrene. *Nanoscale Res. Lett.* **2013**, *8*, 414. [CrossRef]
50. Esteban, R.M.; Schütte, K.; Brandt, P.; Marquardt, D.; Meyer, H.; Beckert, F.; Mülhaupt, R.; Kölling, H.; Janiak, C. Iridium@graphene composite nanomaterials synthesized in ionic liquid as re-usable catalysts for solvent-free hydrogenation of benzene and cyclohexene. *Nano-Struct. Nano-Objects* **2015**, *2*, 11–18. [CrossRef]
51. Wang, W.; Zhu, Y. Shape-controlled synthesis of zinc oxide by microwave heating using an imidazolium salt. *Inorg. Chem. Commun.* **2004**, *7*, 1003–1005. [CrossRef]
52. Wang, J.; Cao, J.; Fang, B.; Lu, P.; Deng, S.; Wang, H. Synthesis and characterization of multipod, flower-like, and shuttle-like Zno frameworks in ionic liquids. *Mate. Lett.* **2005**, *59*, 1405–1408. [CrossRef]

53. Rabieh, S.; Bagheri, M.; Heydari, M.; Badiei, E. Microwave assisted synthesis of ZnO nanoparticles in ionic liquid [Bmim]Cl and their photocatalytic investigation. *Mater. Sci. Semicond. Process.* **2014**, *26*, 244–250. [CrossRef]
54. Esmaili, M.; Habibi-Yangjeh, A. Preparation and characterization of ZnO nanocrystallines in the presence of an ionic liquid using microwave irradiation and photocatalytic activity. *J. Iran. Chem. Soc.* **2010**, *7*, S70–S82. [CrossRef]
55. da Trindade, L.G.; Minervino, G.B.; Trench, A.B.; Carvalho, M.H.; Assis, M.; Li, M.S.; de Oliveira, A.J.A.; Pereira, E.C.; Mazzo, T.M.; Longo, E. Influence of ionic liquid on the photoelectrochemical properties of ZnO particles. *Ceram. Int.* **2018**, *44*, 10393–10401. [CrossRef]
56. Shrestha, N.K.; Macak, J.M.; Schmidt-Stein, F.; Hahn, R.; Mierke, C.T.; Fabry, B.; Schmuki, P. Magnetically guided titania nanotubes for site-selective photocatalysis and drug release. *Angew. Chem. Int. Ed. Engl.* **2010**, *121*, 969–972.
57. Chen, C.; Hu, X.; Hu, P.; Qiao, Y.; Qie, L.; Huang, Y. Ionic-liquid-assisted synthesis of self-assembled TiO_2-B nanosheets under microwave irradiation and their enhanced lithium storage properties. *Eur. J. Inorg. Chem.* **2013**, *30*, 5320–5328. [CrossRef]
58. Li, F.; Wang, X.; Zhao, Y.; Liu, J.; Hao, Y.; Liu, R.; Zhao, D. Ionic-liquid-assisted synthesis of high-visible-light-activated N–B–F-tri-doped mesoporous TiO_2, via a microwave route. *Appl. Catal. B* **2014**, *144*, 442–453. [CrossRef]
59. Zhang, D.; Li, G.; Yang, X.; Yu, J. A micrometer-size TiO_2 single-crystal photocatalyst with remarkable 80% level of reactive facets. *Chem. Commun.* **2009**, *29*, 4381–4383. [CrossRef] [PubMed]
60. Chen, Y.; Li, W.; Wang, J.; Gan, Y.; Liu, L.; Ju, M. Microwave-assisted ionic liquid synthesis of Ti^{3+} self-doped TiO_2 hollow nanocrystals with enhanced visible-light photoactivity. *Appl. Catal. B* **2016**, *191*, 94–105. [CrossRef]
61. Yang, C.; Wang, Z.; Lin, T.; Yin, H.; Lü, X.; Wan, D.; Xu, T.; Zheng, C.; Lin, J.; Huang, F.; et al. Core-shell nanostructured "black" rutile titania as excellent catalyst for hydrogen production enhanced by sulfur doping. *J. Am. Chem. Soc.* **2013**, *135*, 17831–17838. [CrossRef]
62. Xiao, S.; Zhu, W.; Liu, P.; Liu, F.; Dai, W.; Zhang, D.; Chen, W.; Li, H. CNTs threaded (001) exposed TiO_2 with high activity in photocatalytic NO oxidation. *Nanoscale* **2016**, *8*, 2899–2907. [CrossRef] [PubMed]
63. Musa, A.O.; Akomolafe, T.; Carter, M.J. Production of cuprous oxide, a solar cell material, by thermal oxidation and a study of its physical and electrical properties. *Sol. Energy Mater. Sol. Cells* **1998**, *51*, 305–316. [CrossRef]
64. Wang, W.; Zhu, Y.; Cheng, G.; Huang, Y. Microwave-assisted synthesis of cupric oxide nanosheets and nanowhiskers. *Mater. Lett.* **2006**, *60*, 609–612. [CrossRef]
65. Xia, J.; Li, H.; Luo, Z.; Shi, H.; Wang, K.; Shu, H.; Yan, Y. Microwave-assisted synthesis of flower-like and leaf-like CuO nanostructures via room-temperature ionic liquids. *J. Phys. Chem. Solids* **2009**, *70*, 1461–1464. [CrossRef]
66. Zhang, M.; Xu, X.; Zhang, M. Microwave-assisted synthesis and characterization of CuO nanocrystals. *J. Dispersion Sci. Technol.* **2008**, *29*, 508–513. [CrossRef]
67. Ayesh, A.I.; Abu-Hani, A.F.S.; Mahmoud, S.T.; Haik, Y. Selective H_2S sensor based on CuO nanoparticles embedded in organic membranes. *Sens. Actuators B* **2016**, *231*, 593–600. [CrossRef]
68. Zhu, L.; Chen, Y.; Sun, Y.; Cui, Y.; Liang, M.; Zhao, J.; Li, N. Phase-manipulable synthesis of Cu-based nanomaterials using ionic liquid 1-butyl-3-methyl-imidazole tetrafluoroborate. *Cryst. Res. Technol.* **2010**, *45*, 398–404. [CrossRef]
69. Li, S.; Guo, X.; Wang, Y.; Huang, F.; Shen, Y.; Wang, X.; Xie, A. Rapid synthesis of flower-like Cu_2O architectures in ionic liquids by the assistance of microwave irradiation with high photochemical activity. *Dalton Trans.* **2011**, *40*, 6745–6750. [CrossRef]
70. Bowers, J.; Butts, C.P.; Martin, P.J.; Vergara-Gutierrez, M.; Heenan, R.K. Aggregation behavior of aqueous solutions of ionic liquids. *Langmuir* **2004**, *20*, 2191–2198. [CrossRef]
71. Ahmed, F.; Arshi, N.; Anwar, M.S.; Danish, R.; Koo, B. Quantum-confinement induced enhancement in photocatalytic properties of iron oxide nanoparticles prepared by Ionic liquid. *Ceram. Int.* **2014**, *40*, 15743–15751. [CrossRef]
72. Cao, S.; Zhu, Y. Iron oxide hollow spheres: Microwave–hydrothermal ionic liquid preparation, formation mechanism, crystal phase and morphology control and properties. *Acta Mater.* **2009**, *57*, 2154–2165. [CrossRef]

73. Yin, S.; Luo, Z.; Xia, J.; Li, H. Microwave-assisted synthesis of Fe_3O_4 nanorods and nanowires in an ionic liquid. *J. Phys. Chem. Solids* **2010**, *71*, 1785–1788. [CrossRef]
74. Jadhav, A.H.; Lim, A.C.; Thorat, G.M.; Jadhav, H.; Seo, J.G. Green solvent ionic liquids: Structural directing pioneers for microwave-assisted synthesis of controlled MgO nanostructures. *RSC Adv.* **2016**, *6*, 31675–31686. [CrossRef]
75. Qin, B.; Zhang, H.; Diemant, T.; Geiger, D.; Raccichini, R.; Behm, R.J.; Kaiser, U.; Varzi, A.; Passerini, S. Ultrafast ionic liquid-assisted microwave synthesis of SnO microflowers and their superior sodium-Ion storage performance. *ACS Appl. Mater. Interfaces* **2017**, *9*, 26797–26804. [CrossRef] [PubMed]
76. Cybinska, J.; Lorbeer, C.; Mudring, A.V. Ionic liquid assisted microwave synthesis route towards color-tunable luminescence of lanthanide-doped $BiPO_4$. *J. Lumin.* **2016**, *170*, 641–647. [CrossRef]
77. Siramdas, R.; Mclaurin, E.J. InP nanocrystals with color-tunable luminescence by microwave-assisted ionic-liquid etching. *Chem. Mater.* **2017**, *29*, 2101–2109. [CrossRef]
78. Olchowka, J.; Suta, M.; Wickleder, C. Green synthesis of small nanoparticles of A_2SiF_6 (A = Li - Cs) using ionic liquids as solvent and fluorine source: A novel simple approach without HF. *Chemistry* **2017**, *23*, 12092–12095. [CrossRef] [PubMed]
79. Alammar, T.; Hamm, I.; Grasmik, V.; Wark, M.; Mudring, A.V. Microwave-assisted synthesis of perovskite $SrSnO_3$ nanocrystals in ionic liquids for photocatalytic applications. *Inorg. Chem.* **2017**, *56*, 6920–6932. [CrossRef]
80. Nguyen, V.H.; Shim, J.J. Microwave-assisted synthesis of porous nickel cobaltite with different morphologies in ionic liquid and their application in supercapacitors. *Mater. Chem. Phys.* **2016**, *176*, 6–11. [CrossRef]
81. Schmitz, A.; Schütte, K.; Ilievski, V.; Barthel, J.; Burk, L.; Mülhaupt, R.; Yue, J.; Smarsly, B.; Janiak, C. Synthesis of metal-fluoride nanoparticles supported on thermally reduced graphite oxide. *Beilstein J. Nanotechnol.* **2017**, *8*, 2474–2483. [CrossRef] [PubMed]
82. Schaumann, J.; Loor, M.; Ünal, D.; Mudring, A.; Heimann, S.; Hagemann, U.; Schulz, S.; Maculewicz, F.; Schierning, G. Improving the zT value of thermoelectrics by nanostructuring: Tuning the nanoparticle morphology of Sb_2Te_3 by using ionic liquids. *Dalton Trans.* **2017**, *46*, 656–668. [CrossRef] [PubMed]
83. Miao, Y.; Lian, Z.; Huo, Y.; Li, H. Microwave-assisted ionothermal synthesis of hierarchical microcube-like BiOBr with enhanced photocatalytic activity. *Chin. J. Catal.* **2018**, *39*, 1411–1417. [CrossRef]
84. Alammar, T.; Slowing, I.I.; Anderegg, J.; Mudring, A.V. Ionic liquid assisted microwave synthesis of solid solutions of $Sr_{1-x}Ba_xSnO_3$ perovskite for photocatalytic applications. *ChemSusChem* **2017**, *10*, 3387–3401. [CrossRef] [PubMed]
85. Li, T.; Tang, D.; Li, C. Microwave-assisted synthesis of cobalt phosphide using ionic liquid as Co and P dual-source for hydrogen evolution reaction. *Electrochim. Acta* **2019**, *295*, 1027–1033. [CrossRef]
86. Du, C.; Li, J.; Huang, X. Microwave-assisted ionothermal synthesis of $SnSe_x$ nanodots: A facile precursor approach towards $SnSe_2$ nanodots/graphene nanocomposite. *RSC Adv.* **2016**, *6*, 9835–9842. [CrossRef]
87. Xia, J.; Ge, Y.; Zhao, D.; Di, J.; Ji, M.; Yin, S.; Li, H.; Chen, R. Microwave-assisted synthesis of few-layered MoS_2/BiOBr hollow microspheres with superior visible-light-response photocatalytic activity for ciprofloxacin removal. *CrystEngComm* **2015**, *17*, 3645–3651. [CrossRef]
88. Guerrero-Sánchez, C.; Saldívar, E.; Hernández, M.; Arturo, J. A practical, systematic approach for the scaling-up and modeling of industrial copolymerization reactors. *Polym. React. Eng.* **2003**, *11*, 457–506. [CrossRef]
89. Nising, P.; Zeilmann, T.; Meyer, T. On the degradation and stabilization of poly(methyl methacrylate) in a continuous process. *Chem. Eng. Technol.* **2003**, *26*, 599–604. [CrossRef]
90. Mallakpour, S.; Rafiee, Z. New developments in polymer science and technology using combination of ionic liquids and microwave irradiation. *Prog. Polym. Sci.* **2011**, *36*, 1754–1765. [CrossRef]
91. Luxenhofer, R.; Jordan, R. Click chemistry with poly(2-oxazoline)s. *Macromolecules* **2006**, *39*, 3509–3516. [CrossRef]
92. Rossegger, E.; Schenk, V.; Wiesbrock, F. Design strategies for functionalized poly(2-oxazoline)s and derived materials. *Polymers* **2013**, *5*, 956–1011. [CrossRef]
93. Luef, K.P.; Petit, C.; Ottersböck, B.; Ehrenfeld, F.; Grassl, B.; Reynaud, S.; Wiesbrock, F. UV-mediated thiol-ene click reactions for the synthesis of drug-loadable and degradable gels based on copoly(2-oxazoline)s. *Eur. Polym. J.* **2016**, *88*, 701–712. [CrossRef] [PubMed]

94. Hoogenboom, R.; Fijten, M.W.M.; Brändli, C.; Schroer, J.; Schubert, U.S. Automated parallel temperature optimization and determination of activation energy for the living cationic polymerization of 2-ethyl-2-oxazoline. *Macromol. Rapid Commun.* **2013**, *24*, 98–103. [CrossRef]
95. Park, J.S.; Kataoka, K. Precise control of lower critical solution temperature of thermosensitive poly(2-isopropyl-2-oxazoline) via gradient copolymerization with 2-ethyl-2-oxazoline as a hydrophilic comonomer. *Macromolecules* **2006**, *39*, 6622–6630. [CrossRef]
96. Petit, C.; Grassl, B.; Mignard, E. Living cationic ring-opening polymerization of 2-ethyl-2-oxazoline following sustainable concepts: Microwave-assisted and droplet-based millifluidic processes in an ionic liquid medium. *Polym. Chem.* **2017**, *8*, 5910–5917. [CrossRef]
97. Guerrero-Sanchez, C.; Hoogenboom, R.; Schubert, U.S. Fast and "green" living cationic ring opening polymerization of 2-ethyl-2-oxazoline in ionic liquids under microwave irradiation. *Chem. Commun.* **2006**, *28*, 3797–3799. [CrossRef]
98. Guerrero-Sanchez, C.; Lobert, M.; Hoogenboom, R. Microwave-assisted homogeneous polymerizations in water-soluble ionic liquids: An alternative and green approach for polymer synthesis. *Macromol. Rapid Commun.* **2007**, *28*, 456–464. [CrossRef]
99. Tan, Y.; Cai, S.; Liao, L.; Wang, Q.; Liu, L. Microwave-assisted ring-opening polymerization of ε-Caprolactone in the presence of ionic liquid. *Polym. J.* **2009**, *41*, 849–854. [CrossRef]
100. Liao, L.; Zhang, C.; Gong, S. Microwave-assisted ring-opening polymerization of trimethylene carbonate in the presence of ionic liquid. *J. Polym. Sci.* **2007**, *45*, 5857–5863. [CrossRef]
101. Mallakpour, S.; Dinari, M. A study of the ionic liquid mediated microwave heating for the synthesis of new thermally stable and optically active aromatic polyamides under green procedure. *Macromol. Res.* **2010**, *18*, 129–136. [CrossRef]
102. Mallakpour, S.; Zadehnazari, A. Fast synthesis, using microwave induction heating in ionic liquid and characterization of optically active aromatic polyamides. *J. Macromol. Sci. A* **2009**, *46*, 783–789. [CrossRef]
103. Mallakpour, S.; Zadehnazari, A. Microwave-assisted synthesis and morphological characterization of ciral poly(amide-imide) nanostructures in molten ionic liquid salt. *Adv. Polym. Technol.* **2013**, *32*, 21333. [CrossRef]
104. Mallakpour, S.; Rafiee, Z. Use of ionic liquid and microwave irradiation as a convenient, rapid and eco-friendly method for synthesis of novel optically active and thermally stable aromatic polyamides containing N-phthaloyl-l-alanine pendent group. *Polym. Degrad. Stab.* **2008**, *93*, 753–759. [CrossRef]
105. Schmidt-Naake, G.; Woecht, I.; Schmalfub, A.; Glück, T.V. polymer reaction engineering free radical polymerization in ionic liquids - Solvent influence of new dimension. *Macromol. Symp.* **2009**, *275*, 204–218. [CrossRef]
106. Glück, T.; Woecht, I.; Schmalfub, A.; Schmidt-Naake, G. Modification of the solvent influence on the free radical polymerization in ionic liquids by microwave irradiation. *Macromol. Symp.* **2009**, *275*, 230–241. [CrossRef]
107. Wei, X.; Chang, G.; Li, J.; Wang, F.; Cui, L.; Fu, T.; Kong, L. Preparation of pH- and salinity-responsive cellulose copolymer in ionic liquid. *J. Polym. Res.* **2014**, *21*, 535. [CrossRef]
108. Lin, C.; Zhan, H.; Liu, M.; Fu, S.; Huang, L. Rapid homogeneous preparation of cellulose graft copolymer in BMIMCl under microwave irradiation. *J. Appl. Polym. Sci.* **2010**, *118*, 399–404. [CrossRef]
109. Rafiemanzelat, F.; Abdollahi, E. Rapid synthesis of new block copolyurethanes derived froml-leucine cyclodipeptide in reusable molten ammonium salts: Novel and efficient green media for the synthesis of new hydrolysable and biodegradable copolyurethanes. *Amino Acids* **2012**, *42*, 2177–2186. [CrossRef]
110. Döbbelin, M.; Azcune, I.; Bedu, M.; Luzuriaga, A.R.; Genua, A.; Jovanovski, V.; Cabañero, G.; Odriozola, I. Synthesis of pyrrolidinium-based poly(ionic liquid) electrolytes with poly(ethylene glycol) side chains. *Chem. Mater.* **2012**, *24*, 1583–1590. [CrossRef]
111. Tang, J.; Radosz, M.; Shen, Y. Poly(ionic liquid)s as optically transparent microwave-absorbing materials. *Macromolecules* **2008**, *41*, 493–496. [CrossRef]
112. Dong, Y.; Yin, J.; Zhao, X. Microwave-synthesized poly(ionic liquid) particles: A new material with high electrorheological activity. *J. Mater. Chem. A* **2014**, *2*, 9812–9819. [CrossRef]
113. Dong, Y.; Yin, J.; Yuan, J.; Zhao, X. Microwave-assisted synthesis and high-performance anhydrous electrorheological characteristic of monodisperse poly(ionic liquid) particles with different size of cation/anion parts. *Ploymer* **2016**, *97*, 408–471. [CrossRef]

114. Liu, G.; Su, P.; Zhou, L.; Yang, Y. Microwave-assisted preparation of poly(ionic liquid)-modified polystyrene magnetic nanospheres for phthalate esters extraction from beverages. *J. Sep. Sci.* **2017**, *40*, 2603–2611. [CrossRef]
115. Zhang, R.; Su, P.; Yang, L.; Yang, Y. Microwave-assisted preparation of poly(ionic liquids)-modified magnetic nanoparticles for pesticide extraction. *J. Sep. Sci.* **2014**, *37*, 1503–1510. [CrossRef]
116. Joo, S.H.; Choi, S.J.; Oh, I.; Kwak, J.; Liu, Z.; Terasaki, O.; Ryoo, R. Ordered nanoporous arrays of carbon supporting high dispersions of platinum nanoparticles. *Nature* **2001**, *412*, 169–172. [CrossRef] [PubMed]
117. Su, D.; Perathoner, S.; Centi, G. Nanocarbons for the development of advanced catalysts. *Chem. Rev.* **2013**, *113*, 5782–5816. [CrossRef] [PubMed]
118. Tunckol, M.; Durand, J.; Serp, P. Carbon nanomaterial-ionic liquid hybrids. *Carbon* **2012**, *50*, 4303–4334. [CrossRef]
119. Paraknowitsch, J.P.; Thomas, A. Functional carbon materials from ionic liquid precursors. *Macromol. Chem. Phys.* **2012**, *213*, 1132–1145. [CrossRef]
120. Paraknowitsch, J.P.; Zhang, Y.; Thomas, A. Synthesis of mesoporous composite materials of nitrogen-doped carbon and silica using a reactive surfactant approach. *J. Mater. Chem.* **2011**, *21*, 15537–15543. [CrossRef]
121. Xie, Z.; White, R.J.; Weber, J.; Taubert, A.; Titirici, M.M. Hierarchical porous carbonaceous materials via ionothermal carbonization of carbohydrates. *J. Mater. Chem.* **2011**, *21*, 7434–7442. [CrossRef]
122. Lee, J.; Wang, X.; Luo, H.; Baker, G.A.; Dai, S. Facile ionothermal synthesis of microporous and mesoporous carbons from task specific ionic liquids. *J. Am. Chem. Soc.* **2009**, *131*, 4596–4597. [CrossRef]
123. Georgakilas, V.; Perman, J.A.; Tucek, J.; Zboril, R. Broad family of carbon nanoallotropes: Classification, chemistry, and applications of fullerenes, carbondots, nanotubes, graphene, nanodiamonds, and combined superstructures. *Chem. Rev.* **2015**, *115*, 4744–4822. [CrossRef]
124. Sofia, P.; Emilio, P.; Eugenia, M.F. Graphene and carbon quantum dot-based materials in photovoltaic devices: From synthesis to applications. *Nanomaterials* **2016**, *6*, 157.
125. Pham-Truong, T.N.; Petenzi, T.; Ranjan, C.; Randriamahazaka, H.; Ghilane, J. Microwave assisted synthesis of carbon dots in ionic liquid as metal free catalyst for highly selective production of hydrogen peroxide. *Carbon* **2018**, *130*, 544–552. [CrossRef]
126. Jeong, Y.; Moon, K.; Jeong, S.; Koh, W.; Lee, K. Converting waste papers to fluorescent carbon dots in the recycling process without loss of ionic liquids and bio-imaging applications. *ACS Sustainable Chem. Eng.* **2018**, *6*, 4510–4515. [CrossRef]
127. Liu, R.; Gao, M.; Zhang, J.; Li, Z.; Chen, J.; Liu, P.; Wu, D. An ionic liquid promoted microwave-hydrothermal route towards highly photoluminescent carbon dots for sensitive and selective detection of iron. *RSC Adv.* **2015**, *5*, 24205–24209. [CrossRef]
128. Liu, X.; Li, T.; Wu, Q.; Yan, X.; Wu, C.; Chen, X.; Zhang, G. Carbon nanodots as a fluorescence sensor for rapid and sensitive detection of Cr(VI) and their multifunctional applications. *Talanta* **2017**, *165*, 216–222. [CrossRef]
129. Safavi, A.; Sedaghati, F.; Shahbaazi, H.; Farjami, E. Facile approach to the synthesis of carbon nanodots and their peroxidase mimetic function in azo dyes degradation. *RSC Adv.* **2012**, *2*, 7367–7370. [CrossRef]
130. Xiao, D.; Yuan, D.; He, H.; Gao, M. Microwave assisted one-step green synthesis of fluorescent carbon nanoparticles from ionic liquids and their application as novel fluorescence probe for quercetin determination. *J. Lumin.* **2013**, *140*, 120–125. [CrossRef]
131. Wei, Y.; Liu, Y.; Li, H.; Zhang, Q.; Kang, Z.; Lee, S.T. Carbon nanoparticle ionic liquid hybrids and their photoluminescence properties. *J. Colloid Sci.* **2011**, *358*, 146–150. [CrossRef]
132. Meng, Y.; Han, W.; Zhang, Z.; Zhu, F.; Zhang, Y.; Wang, D. $LiFePO_4$ particles coated with N-doped carbon membrane. *J. Nanosci. Nanotechnol.* **2017**, *17*, 2000–2005. [CrossRef]
133. Vadahanambi, S.; Chun, H.H.; Jung, K.H.; Park, H. Nitrogen doped holey carbon nano-sheets as anodes in sodium ion battery. *RSC Adv.* **2016**, *6*, 38112–38116. [CrossRef]
134. Beggs, K.M.; Perus, M.D.; Servinis, L.; O'Dell, L.A.; Fox, B.L.; Gengenbach, T.R.; Henderson, L.C. Rapid surface functionalization of carbon fibres using microwave irradiation in an ionic liquid. *RSC Adv.* **2016**, *6*, 32480–32483. [CrossRef]
135. Myriam, B.; Zois, S.; Maurizio, P. Ionic liquids plus microwave irradiation: A general methodology for the retro-functionalization of single-walled carbon nanotubes. *Nanoscale* **2018**, *10*, 15782–15787.

136. Li, S.; Jia, N.; Zhu, J.; Ma, M.; Sun, R. Synthesis of cellulose-calcium silicate nanocomposites in ethanol/water mixed solvents and their characterization. *Carbohydr. Polym.* **2010**, *80*, 270–275. [CrossRef]
137. Jia, N.; Li, S.; Ma, M.; Sun, R.; Zhu, L. Green microwave-assisted synthesis of cellulose/calcium silicate nanocomposites in ionic liquids and recycled ionic liquids. *Carbohydr. Res.* **2011**, *346*, 2970–2974. [CrossRef]
138. Ma, M.; Qing, S.; Li, S.; Zhu, J.; Sun, R. Microwave synthesis of cellulose/CuO nanocomposites in ionic liquid and its thermal transformation to CuO. *Carbohydr. Polym.* **2013**, *91*, 162–168. [CrossRef]
139. Ma, M.; Dong, Y.; Fu, L.; Li, S.; Sun, R. Cellulose/$CaCO_3$ nanocomposites: Microwave ionic liquid synthesis, characterization, and biological activity. *Carbohydr. Polym.* **2013**, *92*, 1669–1676. [CrossRef]
140. Sharma, M.; Bajpai, A. Superabsorbent nanocomposite from sugarcane bagasse, chitin and clay: Synthesis, characterization and swelling behaviour. *Carbohydr. Polym.* **2018**, *193*, 281–288. [CrossRef]
141. Lu, P.; Cheng, F.; Ou, Y.; Lin, M.; Su, L.; Chen, S.; Yao, X.; Liu, D. Rapid fabrication of transparent film directly from wood fibers with microwave-assisted ionic liquids technology. *Carbohydr. Polym.* **2017**, *174*, 330–336. [CrossRef]
142. Bagheri, M.; Rabieh, S. Preparation and characterization of cellulose-ZnO nanocomposite based on ionic liquid ([C_4mim]Cl). *Cellulose* **2013**, *20*, 699–705. [CrossRef]
143. Lin, C.; Zhu, M.; Zhang, T.; Liu, Y.; Lv, Y.; Li, X.; Liu, M. Cellulose/SnS_2 composite with enhanced visible-light photocatalytic activity prepared by microwave-assisted ionic liquid method. *RSC Adv.* **2017**, *7*, 12255–12264. [CrossRef]
144. Deng, F.; Fu, L.; Ma, M. Microwave-assisted rapid synthesis and characterization of CaF_2 particles-filled cellulose nanocomposites in ionic liquid. *Carbohydr. Polym.* **2015**, *121*, 163–168. [CrossRef]
145. Román-Leshkov, Y.; Barrett, C.J.; Liu, Z.Y.; Dumesic, J.A. production of dimethylfuran for liquid fuels from biomass-derived carbohydrates. *Nature* **2007**, *447*, 982–985. [CrossRef] [PubMed]
146. Rosatella, A.A.; Simeonov, S.P.; Frade, R.F.M.; Afonso, C.A.M. 5-Hydroxymethylfurfural (HMF) as a building block platform: Biological properties, synthesis and synthetic applications. *Green Chem.* **2011**, *13*, 754–793. [CrossRef]
147. De, S.; Dutta, S.; Saha, B. Microwave assisted conversion of carbohydrates and biopolymers to 5-hydroxymethylfurfural with aluminium chloride catalyst in water. *Green Chem.* **2011**, *13*, 2859–2868. [CrossRef]
148. Li, C.; Zhang, Z.; Zhao, Z.K. Direct conversion of glucose and cellulose to 5-hydroxymethylfurfural in ionic liquid under microwave irradiation. *Tetrahedron Lett.* **2009**, *50*, 5403–5405. [CrossRef]
149. Qi, X.; Watanabe, M.; Aida, T.M.; Smith, R.L.J. Fast transformation of glucose and di-/polysaccharides into 5-Hydroxymethylfurfural by microwave heating in an ionic liquid/catalyst system. *ChemSusChem.* **2010**, *3*, 1071–1077. [CrossRef]
150. Qu, Y.; Wei, Q.; Li, H.; Oleskowicz-Popiel, P.; Huang, C.; Xu, J. Microwave-assisted conversion of microcrystalline cellulose to 5-hydroxymethylfurfural catalyzed by ionic liquids. *Bioresour. Technol.* **2014**, *162*, 358–364. [CrossRef]
151. Zhang, C.; Cheng, Z.; Fu, Z.; Liu, Y.; Yi, X.; Zheng, A.; Kirk, S.R.; Yin, D. Effective transformation of cellulose to 5-hydroxymethylfurfural catalyzed by fluorine anion-containing ionic liquid modified biochar sulfonic acids in water. *Cellulose* **2017**, *24*, 95–106. [CrossRef]
152. Xing, R.; Subrahmanyam, A.V.; Olcay, H.; Qi, W.; Van Walsum, G.P.; Pendse, H.; Huber, G.W. Production of jet and diesel fuel range alkanes from waste hemicellulose-derived aqueous solutions. *Green Chem.* **2010**, *12*, 1933–1946. [CrossRef]
153. Serrano-Ruiz, J.C.; Campelo, J.; Francavilla, M.; Romero, A.A.; Luque, R.; Menéndez-Vázquez, C.; García, A.B.; García-Suárez, E.J. Efficient microwave-assisted production of furfural from C5 sugars in aqueous media catalysed by Brönsted acidic ionic liquids. *Catal. Sci. Technol.* **2012**, *2*, 1828–1832. [CrossRef]
154. Zhang, L.; Yu, H.; Wang, P.; Dong, H.; Peng, X. Conversion of xylan, d-xylose and lignocellulosic biomass into furfural using $AlCl_3$ as catalyst in ionic liquid. *Bioresour. Technol.* **2013**, *130*, 110–116. [CrossRef]
155. Zhang, Z.; Zhao, Z. Microwave-assisted conversion of lignocellulosic biomass into furans in ionic liquid. *Bioresour. Technol.* **2010**, *101*, 1111–1114. [CrossRef] [PubMed]
156. da Silva Lacerda, V.; López-Sotelo, J.B.; Correa-Guimarães, A.; Hernández-Navarro, S.; Sánchez-Bascones, M.; Navas-Gracia, L.M.; Martín-Ramos, P.; Pérez-Lebeña, E.; Martín-Gil, J. A kinetic study on microwave-assisted conversion of cellulose and lignocellulosic waste into hydroxymethylfurfural/furfural. *Bioresour. Technol.* **2015**, *180*, 88–96. [CrossRef]

157. Chen, J.; Zhang, C.; Li, M.; Chen, J.; Wang, Y.; Zhou, F. Microwave-enhanced sub-critical hydrolysis of rice straw to produce reducing sugar catalyzed by ionic liquid. *J. Mater. Cycles Waste Manag.* **2018**, *20*, 1364–1370. [CrossRef]
158. Hou, X.; Wang, Z.; Sun, J.; Li, M.; Wang, S.; Chen, K.; Gao, Z. A microwave-assisted aqueous ionic liquid pretreatment to enhance enzymatic hydrolysis of Eucalyptus and its mechanism. *Bioresour. Technol.* **2019**, *272*, 99–104. [CrossRef] [PubMed]
159. Nawshad, M.; Zakaria, M.; Mohamad, A. Ionic liquid—A future solvent for the enhanced uses of wood biomass. *Eur. J. Wood Wood Prod.* **2011**, *70*, 125–133.
160. Zhou, Y.; Schattka, J.H.; Antonietti, M. Room-temperature ionic liquids as template to monolithic mesoporous silica with wormlike pores via a sol-gel nanocasting technique. *Nano Lett.* **2004**, *4*, 477–481. [CrossRef]
161. Pang, J.; Luan, Y.; Wang, Q.; Du, J.; Cai, X.; Li, Z. Microwave-assistant synthesis of inorganic particles from ionic liquid precursors. *Colloids Surf. A* **2010**, *360*, 6–12. [CrossRef]

MDPI
St. Alban-Anlage 66
4052 Basel
Switzerland
Tel. +41 61 683 77 34
Fax +41 61 302 89 18
www.mdpi.com

Nanomaterials Editorial Office
E-mail: nanomaterials@mdpi.com
www.mdpi.com/journal/nanomaterials

www.ingramcontent.com/pod-product-compliance
Lightning Source LLC
LaVergne TN
LVHW070659170726
843515LV00004B/449

* 9 7 8 3 0 3 9 4 3 3 0 8 7 *